AF882204

NEMATODE TRADING CARDS

JONATHAN D. EISENBACK

Library of Congress Cataloging-in Publication Data
Names: Eisenback, Jonathan D., 1952-
Title: Nematode Trading Cards Decks 1-3
Description: Mactode Publications [2022]
Series: Decks
Identifiers: ISBN 978-1-893961-47-0 (softcover)
Subjects: LCSH: Nematodes, Nematology

Author and Editor: Jonathan D. Eisenback
Copy Editor: Marilyn L. Eisenback
Art Editor and Designer: Jonathan D. Eisenback

First Edition, 2022
Printed in the United States by Kindle Direct Publishing
and Published by Mactode Publications
Blacksburg, VA 24060
August 6/2022

Copyright 2022 J. D. Eisenback (jon@vt.edu)
All rights reserved

Without limiting the rights under copyright reserved above, no part of this publication may be reproduced, stored in or introduced into a retrieval system, or trans- mitted, in any form, or by any means (electronic, mechanical, photocopying, recording, or otherwise), without the prior written permission of the copyright owner.
Mactode Publication books are available at special discounts when purchased in bulk for educational use. For details, contact the publisher.

Printed and bound in the USA.

Dedicated to

Dr. Joseph Neal Sasser

For his numerous contributions to the advancement of the science of nematology

Preface

These nematode trading cards were a product of the Education Committee of the Society of Nematologists. Each year one deck of 54 cards was given to each participant of the Annual Meeting in their registration package. Deck One was presented at the meeting in Williamsburg, VA; Deck Two was given at the meeting in Albuquerque, NM; and Deck Three was handed out at the meeting in Raleigh, NC.

JDE

Table of Contents

NTC Deck ONE

Nematodes

22. *Ancyclostoma caninum*
23. *Anguina agrostis*
24. *Mesocriconema* spp.
25. *Halicephalobus mephisto*
26. *Sphaerularia bombi*
27. *Pratylenchus* spp.
28. *Mermis nigrescens*
29. *Parelaphostrongylus tenuis*
30. *Trichina spiralis*
31. *Globodera* spp.
32. *Meloidogyne* spp.
33. *Hemicycliophora* spp.

Nematologists

34. Pierre Borellius
35. Emilio Göldi
36. Otto Butschli
37. Ivan Filipjev
38. Charlton Bastian
39. Karl Rudolphi.
40. Carl Llinneaus
41. John Needham
42. Melchor Treub
43. Nathan Cobb
44. Joe Sasser
45. Al Taylor
46. Hedwig Hirschmann
47. Tasso Triantaphyllou
48. George Poinar
49. Jonathan Eisenback

Symptomology

50. Angular leaf spots
51. Galls on roots
52. Lesions on roots

Morphology

53. Bursa
54. Perineal pattern

NTC Deck TWO

Nematologists

1. Alan Bird
2. Brian Kerry
3. Burt Zuckerman
4. Carl Linneaus
5. Felix Dujardin
6. Francesco Palumbo
7. Georg Liebscher
8. Gregor Yeates
9. James Starr
10. Ken Evans
11. M. R. Siddiqi
12. Maria Bravo
13. Michel Luc
14. Randy Gaugler
15. Robert Hooke
16. Rolo Perry
17. Steve Thomas

18. Susan Santos
19. Frida Decraemer
20. Howard Ferris
21. Ken Barker
22. Virginia Ferris

Control

23. *Pochonia chlamydosporia*
24 Tioxazafen
25. Fluensulfone
26. *Burhoderia rinojensis* strain A396
27. Spirotetramat
28. Oxamyl
29. Sodium methyldithiocarbamate
30. Furfural

Habitat

31. Agricultural soils
32. Turfgrass species
33. Seed heads
34. Inside a plant leaf
35. Pulmonary arterial system
36. Sandy beaches

Morphology

37. Female intersex
38. Cyst nematode sperm
39. Odontostyle
40. Stomatostyle

41. Root-knot female ovary
42. Anhydrobiosis
43. Root-knot tetrad chromosomes
44. Rudimentary vulva
45. Perineal pattern

Nematode

46. *Heterodera glycines*
47. *Ditylenchus dipsaci*
48. *Tylocehphalus* spp.
49. *Globodera* spp.
50. *Pratylenchus* spp.
51. *Rotylenchulus reniformis*
52. *Heterodera glycines* juvenile

Symptoms

53. Red ring disease
54. Bronzing of needles

NTC Deck THREE

Nematologists

1. George Atkinson
2. Miles Berkeley
3. Maxime Cornu
4. Casimir Davaine
5. Johannes de Man
6. Alexei Fedchenko
7. Guglielmo Gasparrini

8. Johan Goeze
9. Patrick Manson
10. Otto Müller
11. Richard Owen
12. James Paget
13. Francesco Redi
14. Jan Ritzema-Bos
15. Hermann Schacht
16. Freidrich Schneider
17. Edward Tyson
18. Rudolf Virchow
19. Joseph Leidy
20. Rudolf Leuckart
21. Karl Diesing
22. Frank Lamson-Scribner
23. Ernie Bernard
24. Lawrence Miller
25. Donald Dickson
26. Russ Ingham
27. Robert Robbins
28. Terry Kirkpatrick
29. Diana Wall
30. Ann MacGuidwin
31. John Webster

Habitats

32. Fruit mimicry
33. Human intestine and subcutaneous tissues
34. Mammal skeletal muscles

Nematodes

Behavior

DECK ONE

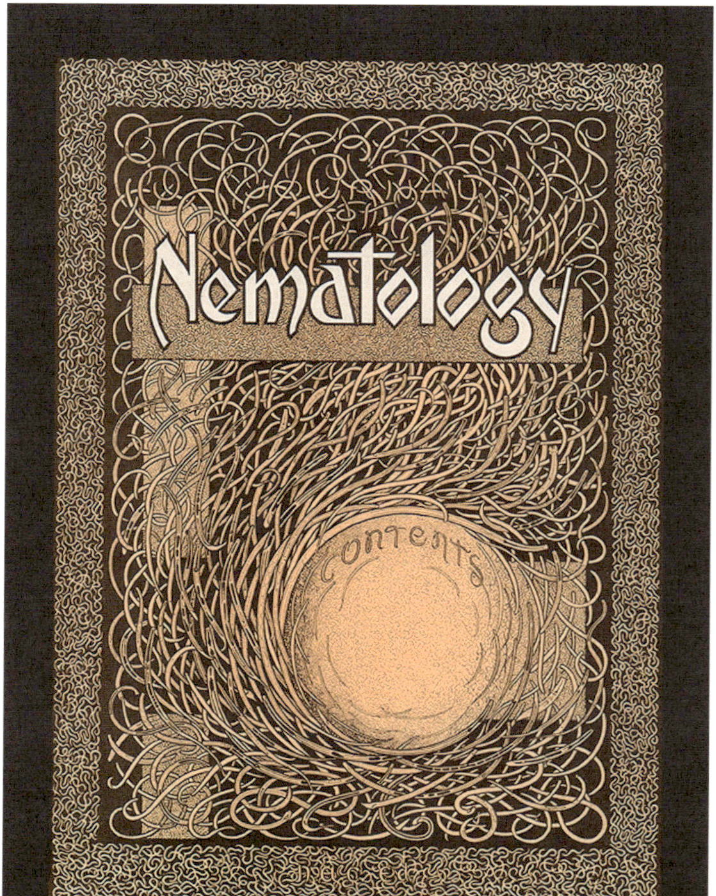
Nematology
CONTENTS

Flowering plant Carnivore

Philcoxia minensis

A carnivorous flowering plant, Philcoxia minensis, catches soil-borne nematodes with sticky knobs on their underground leaves. This genus is made up of three species that are found in areas of the Brazilian Cerrado that have high light levels and nutrient poor soils. This plant produces sticky knobs for capturing nematodes and digests them beneath the surface of the soil to supplement their nitrogen requirements.

① 1

Nematology
CONTENTS

Fungus Sticky networks

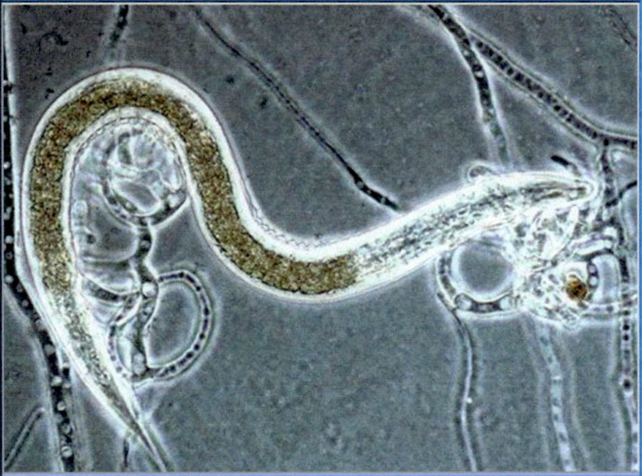

Arthrobotrys oligospora

The first species to be recognized as a nematode-trapping fungus is very common and easy to isolate. This fungus produces a network of sticky traps in response to the presence of nematodes and they secrete a nematotoxin that paralyzes the worms. Fungal hyphae penetrate the culticle of the nematode and digests their contents.

①

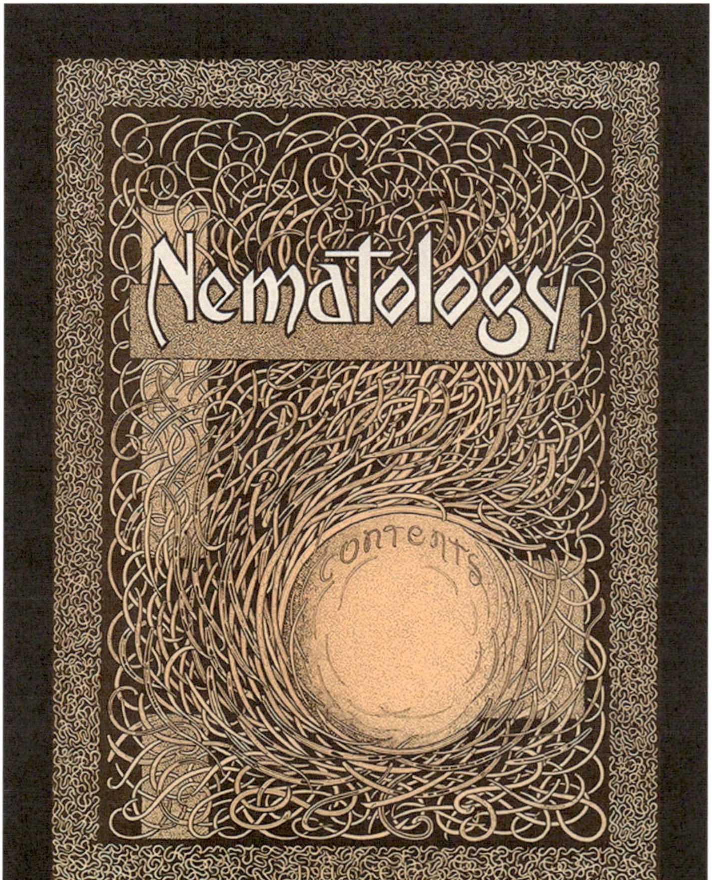
Nematology
CONTENTS

Annelid Predator

Eisenia fetida

Commonly known as redworm, red wriggler worm, branding worm, etc., this species of earthworm thrives in dead organic matter like compost, rotting vegetation, and manure, but rarely in soil. Found in every continent except Antarctica, they are native in Europe. These worms kill nematodes that pass through their digestive system, including those beneficial nematodes that prey on insect larvae.

https://garden.jardinitis.com/productos/compostadores_categoria/vermicompostadores/lombriz-roja-eisenia-foetida-1

① 3

Nematology
Contents

Bacteria Sticky spores

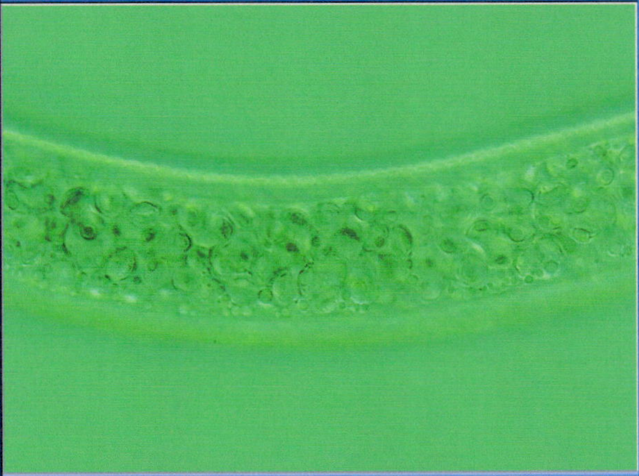

Pasteuria penetrans

Pasteuria penetrans is a gram-positive bacterial pathogen of nematodes that forms endospores. Unfortunately, its fastidious life-cycle has limited its effective use as a biological control agent.

①

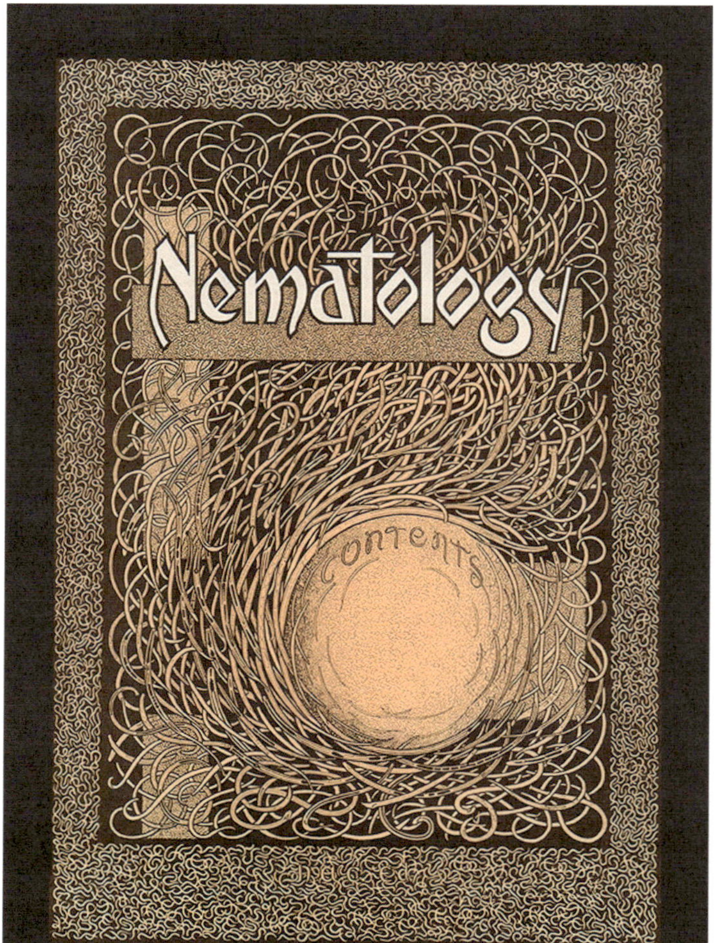
Nematology
CONTENTS

Soybean cyst nematodes

Genetic

Resistance

Soybean cyst nematodes, *Heterodera glycines*, are very pathogenic and cause significant economic losses. SCN resistant varieties are mandatory in the management of this pest. Unfortunately, the use of these varieties have brought about the selection of populations that are able to reproduce on them. These populations are desribed as different "races" or "HG types."

①

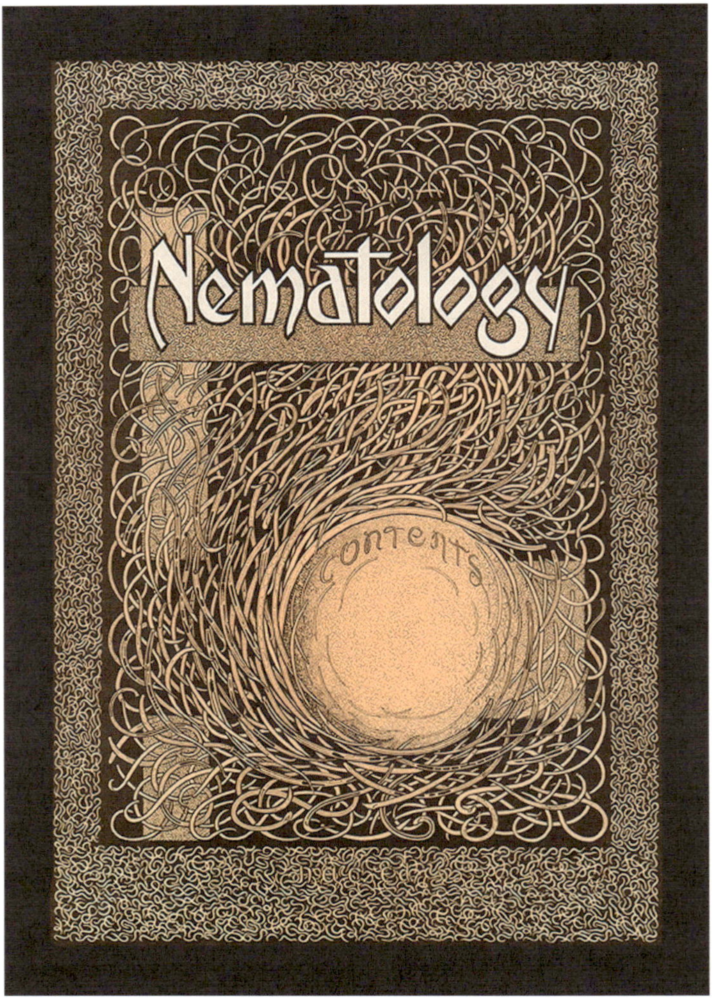
Nematology
Contents

Tardigrade Predator

Tardigrada spp.

Predatory tardigrades feed on soil-borne nematodes, but their affect on the regulation of population density is not very well known. In one study, Macrobiuotus richtersi ate more that 60 nematodes a day which may demonstrate the suppressive nature that has been observed by many independent researchers in the field.

http://www.bbc.com/earth/story/20150313-the-toughest-animals-on-earth

① 6

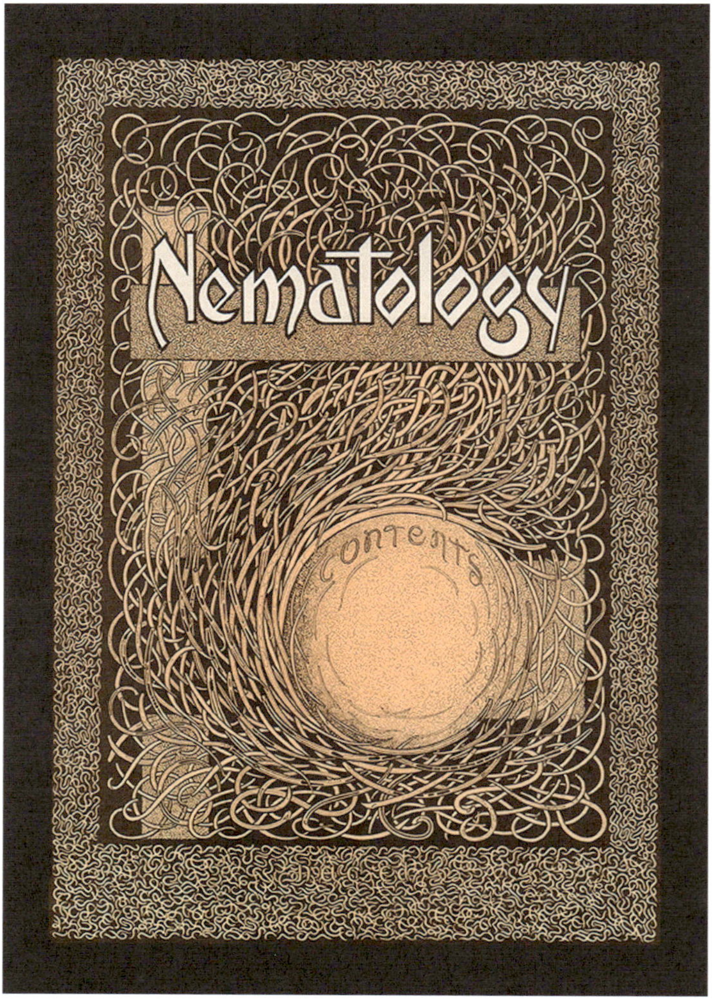
Nematology
CONTENTS

Terr-O-Gas Fumigant

Chemical fumigant

Terr-O-Gas 67 was a restricted-use pre-plant soil fumigant that contained 67% methyl bromide and 33% chloropicrin which was very effective in reducing the numbers of plant-parasitic nematodes in the soil.

① 7

Nematology
Contents

Amoeba Predator

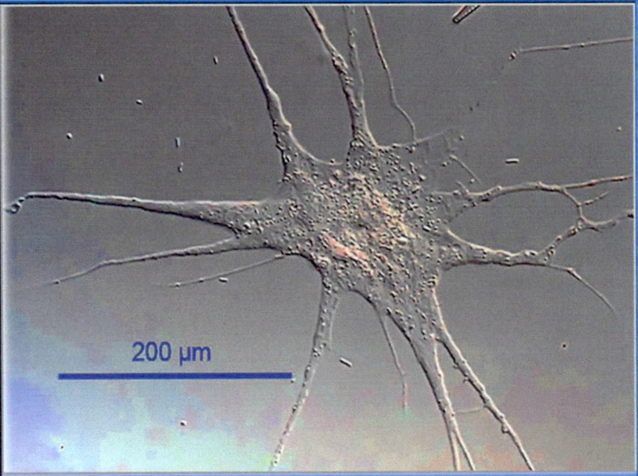

Theratromyxa weberi

Although this amoeba is predatory on plant-parasitic nematodes, greenhouse tests failed to demonstrate that it was useful as a biological control agent.

Martin Kreutz

① 8

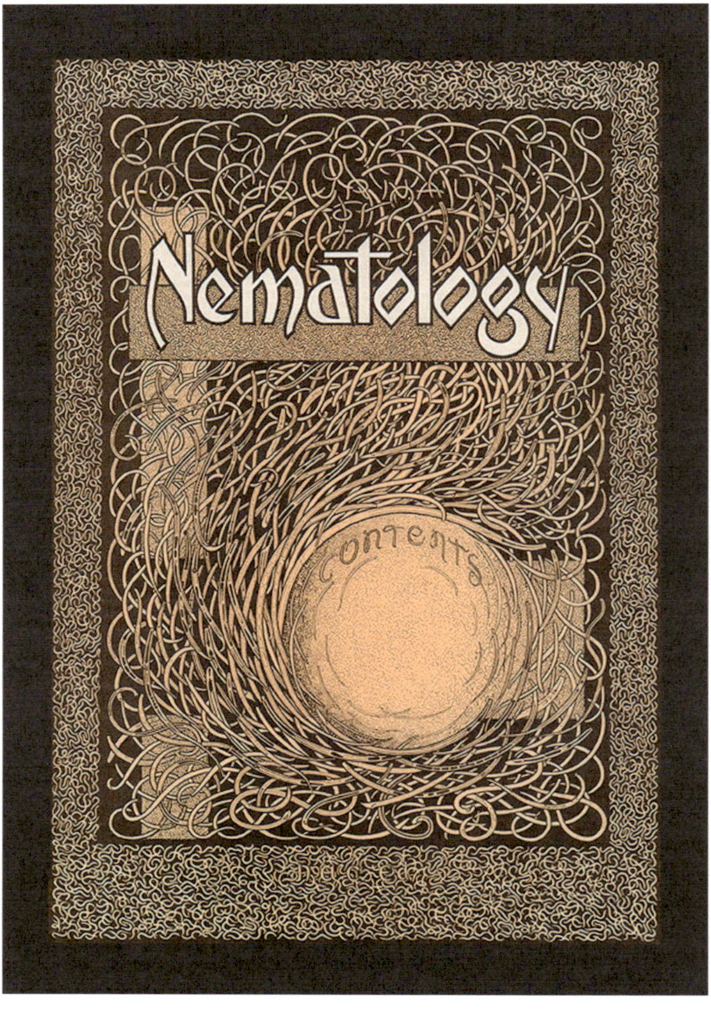
Nematology
Contents

Nematode

Predator

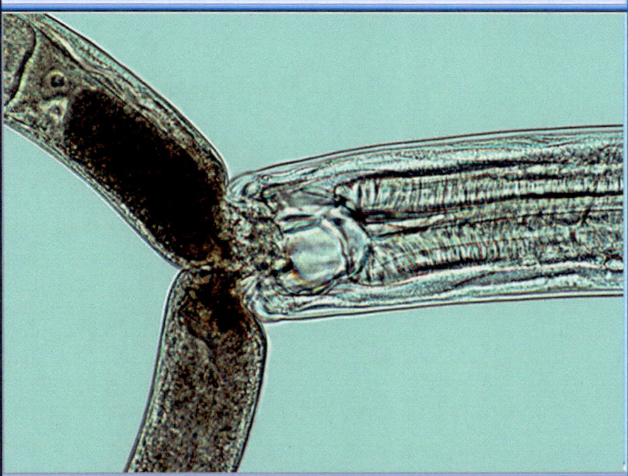

Mononchid nematode

Predatory nematodes from the Mononchid family prey on other nematodes. They use a powerful suction from their esophagus to attach to other nematodes and to suck their body contents into their intestine. A potential biological control agent, but their effective use has been limited by the difficulty in changing the environment of the agricultural field to favor the survival of these predators.

①

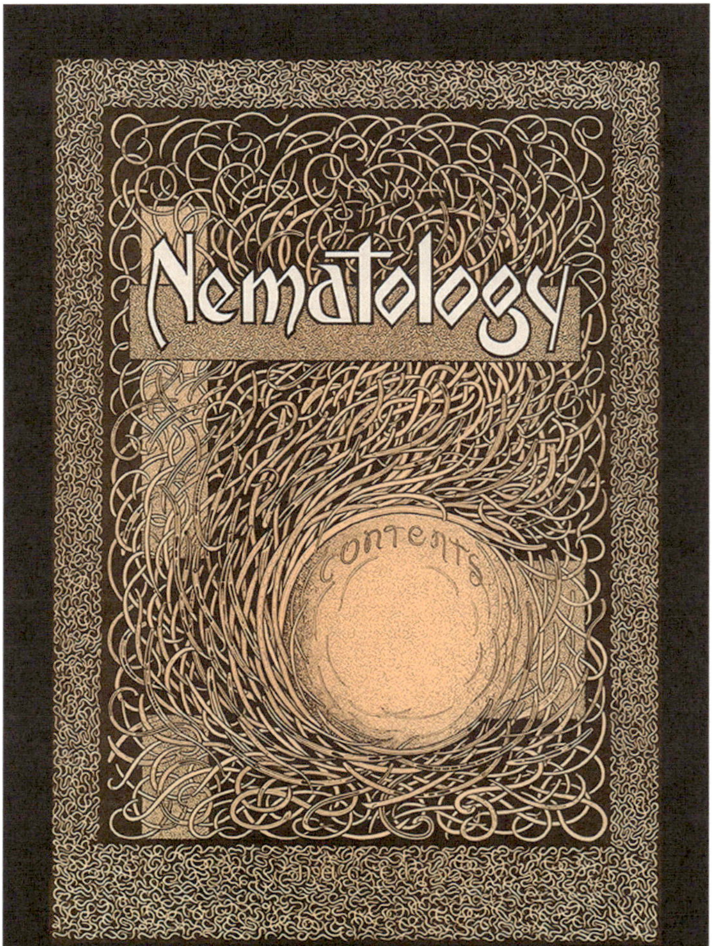
Nematology
CONTENTS

Resistance

Genetic

Root-knot nematode on tomato

Root-knot nematodes, Meloidogyne spp., cause significant economic injury caused by the galls that are produced by plants in response to infection. Some cultivars of plants possess genes that allow them to produce effector proteins that help them to fight off these pathogens, making them resistant to infection. In tomato, this gene is referred to as the "Mi" gene.

①

10

Nematology
Contents

Extreme Artificial

Beer mats

Beer mats (paper coasters) are commonly colonized by Panagrellus redivivus. For many years the very acidic environment of these mats were the only known habitat of this nematode. It has since been found in other fermenting liquids and decomposing wood. Commonly known as the sour paste nematode, it was originally described by Linnaeus from book-binding glue. Aquarists refer to this nematode as microworm.

①

Nematology
Contents

Terrestrial Temperate

Coniferous forest

Coniferous forests are usually better habitats for K-strategists with a long generation time.

① 12

Nematology
contents

Terrestrial & Aquatic Ubiquitous

Planet Earth

Four out of five animals that are alive are nematodes. N.A. Cobb rightly remarked when he said, ..."if all the matter in the universe except the nematodes were swept away, our world would still be dimly recognizable, and if, as disembodied spirits, we could then investigate it, we should find its mountains, hills, vales, rivers, lakes, and oceans represented by a film of nematodes...".

NASA

①

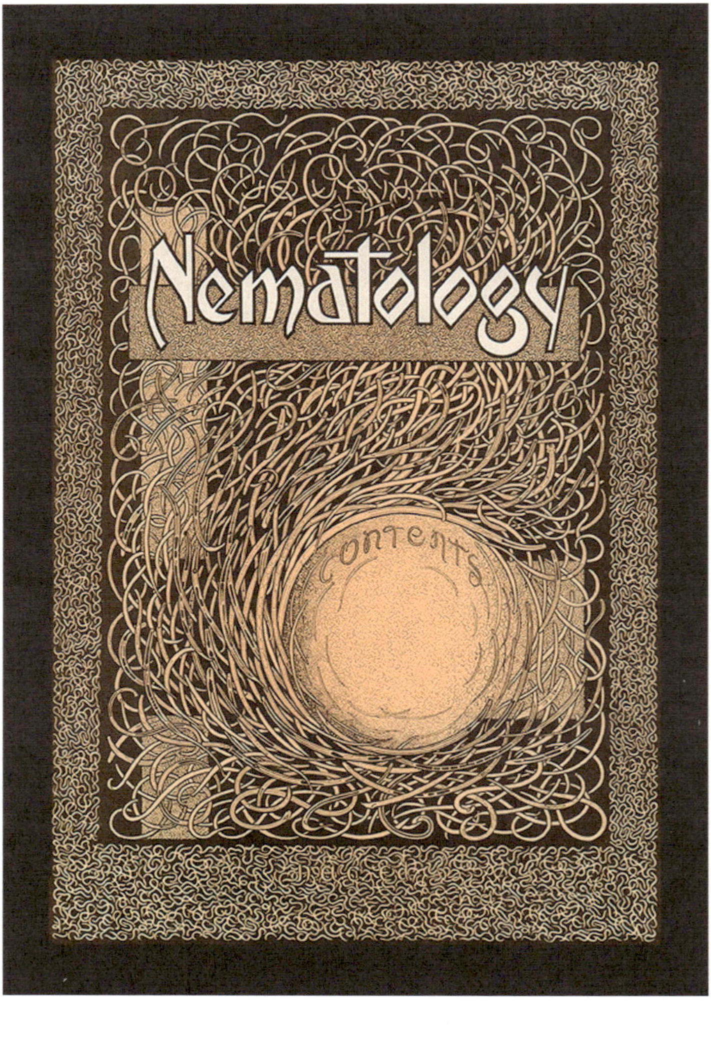
Nematology
Contents

Extreme Subterranian

Deep mine bore holes

Free-living bacteriovors have been found in hot water flowing from mine bore holes more than a mile deep in a gold mine in South Africa.

Gaetan Borgonie

①

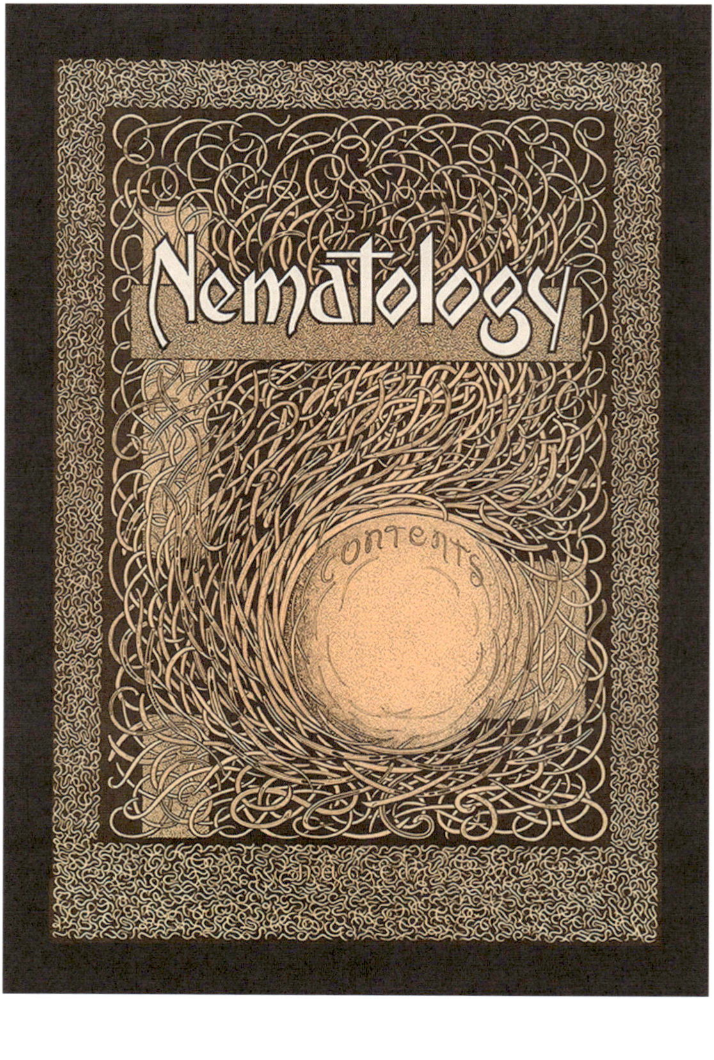
Nematology
Contents

Extreme 2,500 meters deep

Hydrothermal vents

Free-living nematodes are common inhabitants of deep hydrothermal vents often associated with beds of the mussel, Bathymodiolus thermophilus. Seventeen species of nematodes from 14 genera and 11 families have been found, but often one species is the most abundant.

http://faculty.montgomerycollege.edu/gyouth/FP_examples/student_examples/sathya_ramachandran/index.html

①

Nematology
CONTENTS

Extreme 61°C

Hot springs

Nematodes are commonly found in hot springs. Aphelenchoides has been reported surviving 61.3°C, a record, not only for nematodes, but for all metazoans.

① 16

Nematology
CONTENTS

Extreme Frigid

Oil seepage in deep lakes

Nematodes have been found living in the oil seepage of deep, freshwater lakes, i.e. Lake Baikal in Russia. The oil should pollute the lake, but instead it is consumed by bacteria, that are in turn eaten by the nematodes that are living in the thick deposits of hydrocarbons.

A. S. Zakharenko et al

① 17

Nematology
CONTENTS

Extreme Wet/dry/hot/cold

Rough tree bark

Nematodes are very common on the rough bark of trees, however, they are very rare on snooth bark. This habitat is considered to be extreme because when it is raining it is very wet, but after drying out it is very dry. Likewise in the summer it gets very hot, but in winter cold temperatures are common. In order to survive these extremes, the nematodes on bark become anhydrobiotic.

①

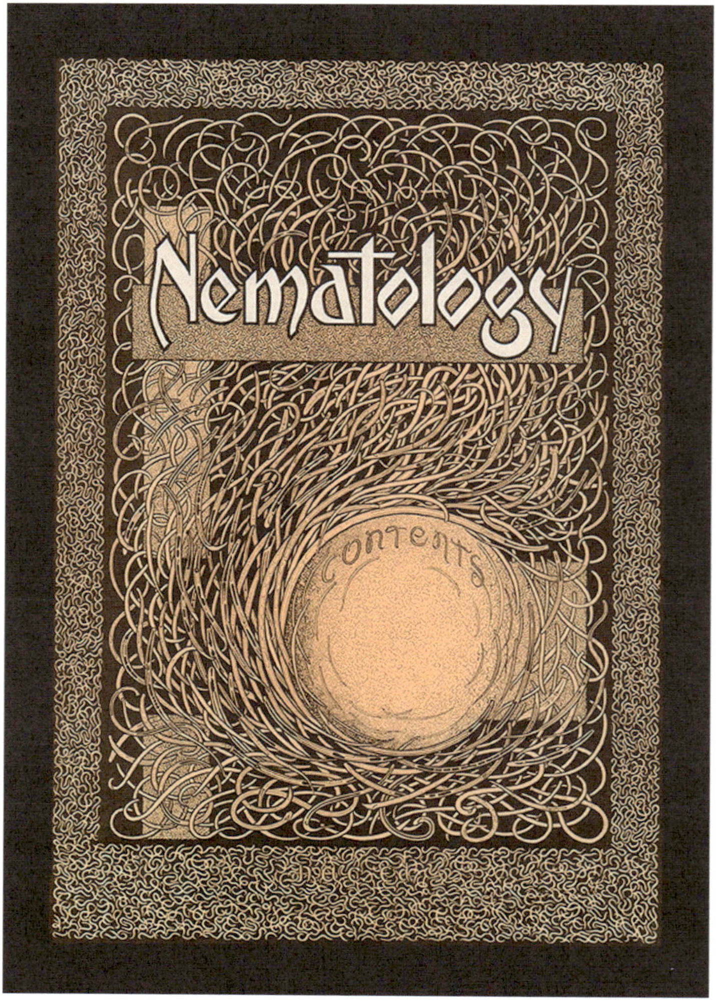
Nematology
CONTENTS

Extreme Subterranean/acidic

Snotites

Nematodes live in whitish hollow mucous-like tubes that hang from the ceiling of caves. These snottites are formed from hydrogen sulfide and oxygen combining to produce sulfuric acid that condenses on the ceiling. Fluid dropping from the snottites have a pH between 0 and 3. It is thought that the low pH may protect the nematodes from predation by mites.

Modified from Kenneth Ingham

① 19

Nematology
CONTENTS

Terrestrial Termperate

Temperate grasslands

Temperate grasslands are usually better habitats for R-strategists which have a high reproductive rate, and are free-living bacteriovors and plant-parasites.

https://www.sophia.org/tutorials/temperate-grasslands--3

①

Nematology
contents

Extreme Ph 2.4-3.4

Vinegar

Vinegar eelworms, Turbatrix aceti, are found in non-pasteurized vinegar, especially if it is raw, unfiltered, and contains "mother". These nematodes are commonly used as a source of food for freshwater fry and even small adult fish that are raised in an aquarium.

http://www.nontoxicspace.com/products/Bragg-Organic-Apple-Cider-Vinegar.html

①

Nematology
CONTENTS

Dog hookworm Animal parasite

Ancyclostoma caninum

The parasite of the small intestine of dogs, but sometimes found in wolves, foxes, and cats. Rarely found in humans. It is restricted to tropical, sub-tropical, and temperate regions of the world, especially in southeast Asian countries such as Malaysia and Sri Lanka.

①

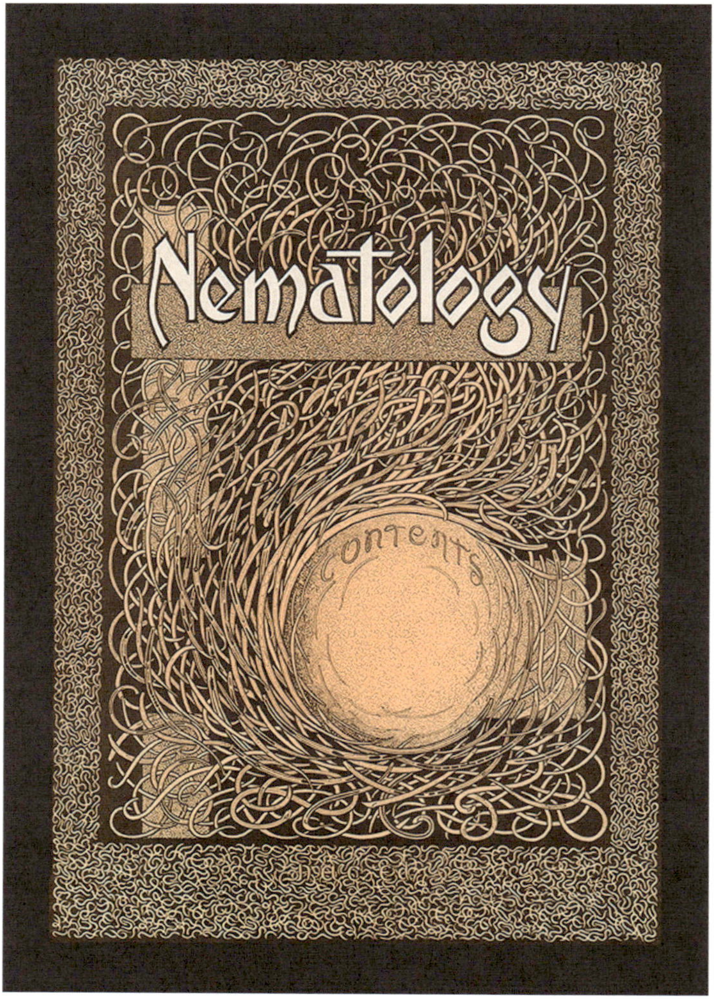
Nematology
Contents

Grass seed gall Aerial parasite

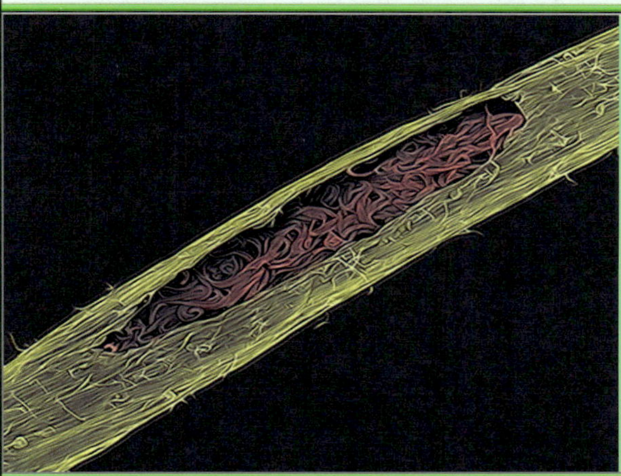

Anguina agrostis

This nematode parasitizes the ovaries of numerous grasses and causes the plant to produce a seed gall which may contain between 500 to 1500 infective juveniles that are in a hydrobiotic condition. It is very important in the Pacific northwest of the U.S. where grass seed production is an important industry. Elsewhere it is of little concern.

①

Nematology
contents

Ring nematode Ectoparasite

***Mesocriconema* spp.**

Ring nematodes: Body short and stout, very sluggish movement, deep annulations, edges smooth, stylet long and thick, anchor-shape knobs, procorpus fused with metacorpus, vulva near tail tip, one ovary, males are rare. Widely distributed on woody plants, turfgrasses, peach, grape, apple, plum, walnut, cherrly, almond, pine, maple, lettuce and numerous other plants.

①

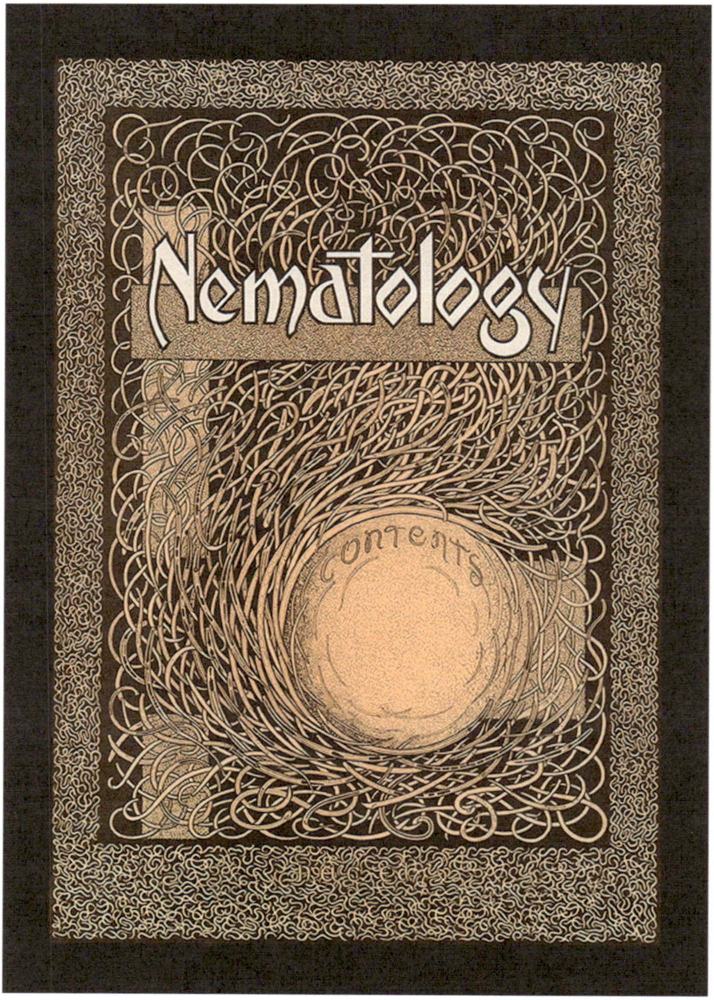
Nematology
CONTENTS

Worm from Hell | Free-living

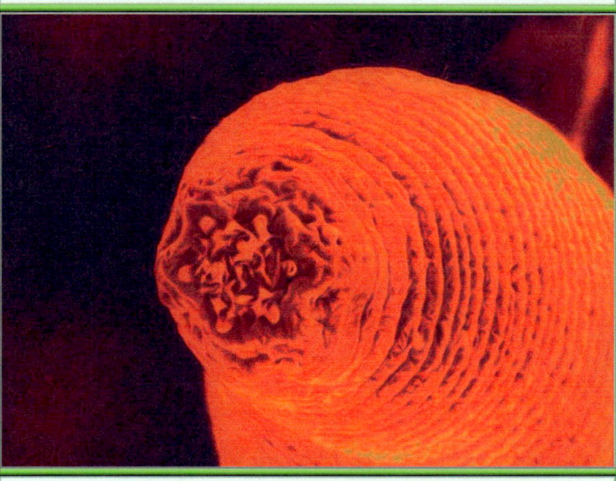

Halicephalobus mephisto

Found a mile below the surface of the earth in water flowing from a bore hole in a gold mine in South Africa.

Gaetan Borgonie

①

Nematology
CONTENTS

Bumble bee nematode Insect parasite

Sphaerularia bombi

The bumble bee nematode parasitizes a queen bumble bee and causes her to forage and feed as normal but prevents her from starting a new colony of bees. Instead she returns to her overwintering site so that the nematodes can emerge and be released back into the soil. The reproductive system turns inside out and greatly increases in size, allowing her to produce tens of thousands of eggs.

http://www.ecoevoblog.com/2012/10/03/the-plight-of-the-bumble-bee-diapause-immunity-and-parasitic-attack/

① 26

Nematology
Contents

Lesion nematode Migratory endoparasite

***Pratylenchus* spp.**

Lesion nematodes: Short, narrow body; broad, flattened head; short, strong stylet with prominent knobs; metacorpus prominent, vulva near tail, one ovary; males common in some species, but rare in others. Widely distributed on woody plants, turfgrasses, peach, grape, apple, plum, walnut, cherry, almond, pine, maple, lettuce, corn, soybean, and numerous other plants.

Uleixh Zunke

①

Nematology
CONTENTS

Grasshopper nematode   Insect parasite

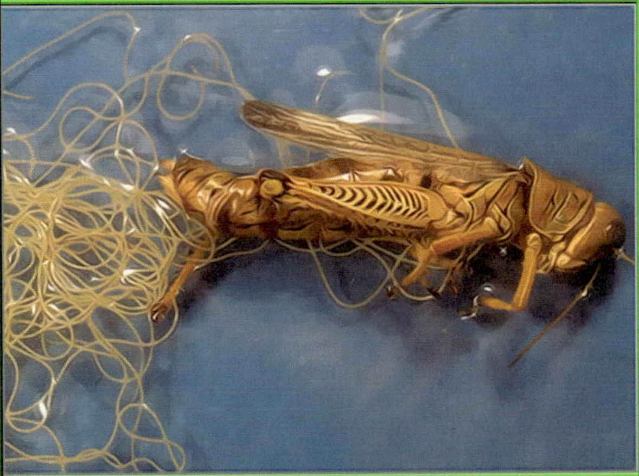

Mermis nigrescens

Juveniles of Mermis nigrescens are parasites of grasshoppers. Adults live in the soil and lay their eggs on plants. They are commonly found in grasslands, meadows, and agriculturals sites.These nematodes are often confused with horsehair worms because they are somewhat similar morphologically and live in the same habitats but belong to separate phyla.

John Capinera, University of Florida

① 28

Nematology
CONTENTS

Brain worm Animal parasite

Parelaphostrongylus tenuis

This nematode causes cerebrospinal nematodiasis in woodland caribou, elk, and moose in the eastern and central North America where white-tailed deer are also common. The adults live in the subdural space of the brain and lay eggs on the dura mater close to small blood vessels. The caribou, elk, and moose are very susceptible to this nematode, but the white-tail deer is asymptomatic.

National Geographic Magazine

①

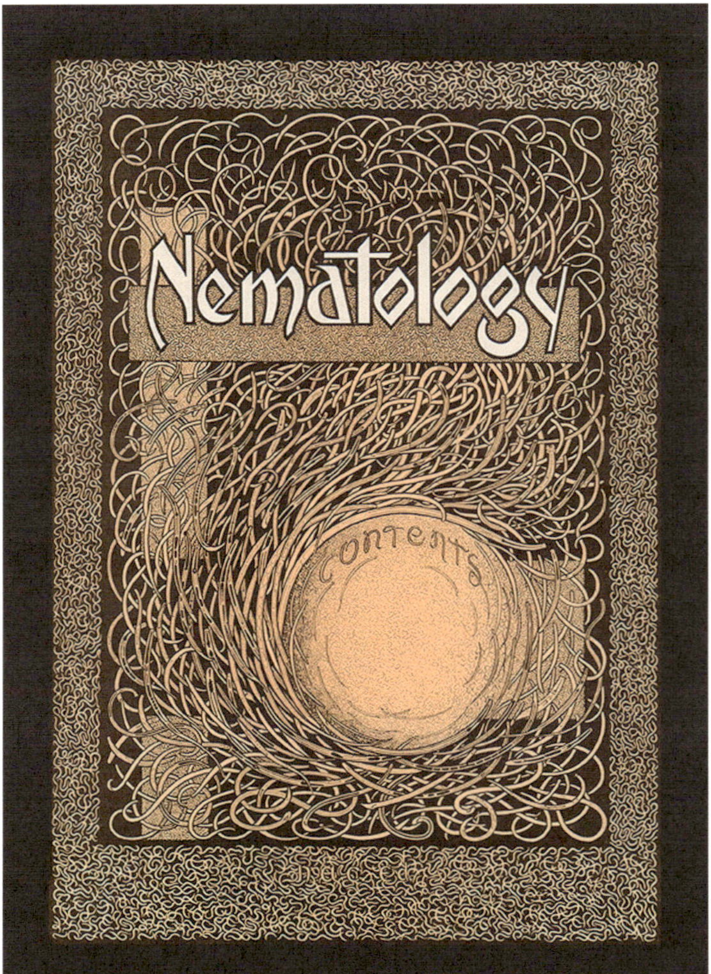
Nematology
CONTENTS

Pork Trichina worm | Animal parasite

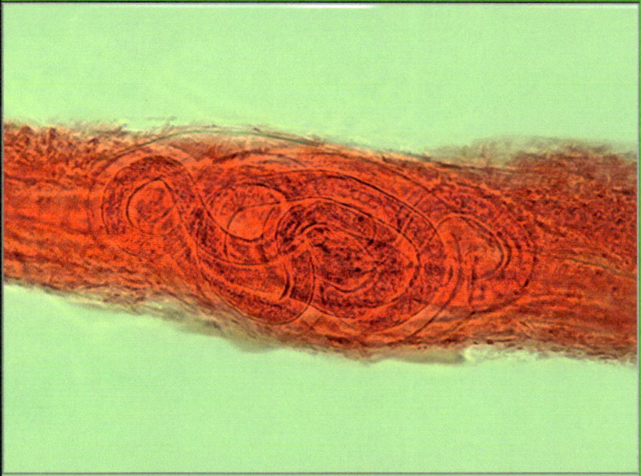

Trichinella spiralis

Infective juveniles become encapsulated within the muscles cells of certain mammals, including pig, rat, dog, and bear. Humans are susceptible if they ingest meat that is not cooked thoroughly. These parasites are common in pork and bear meat and can be fatal if they are ingested in large quantities.

①

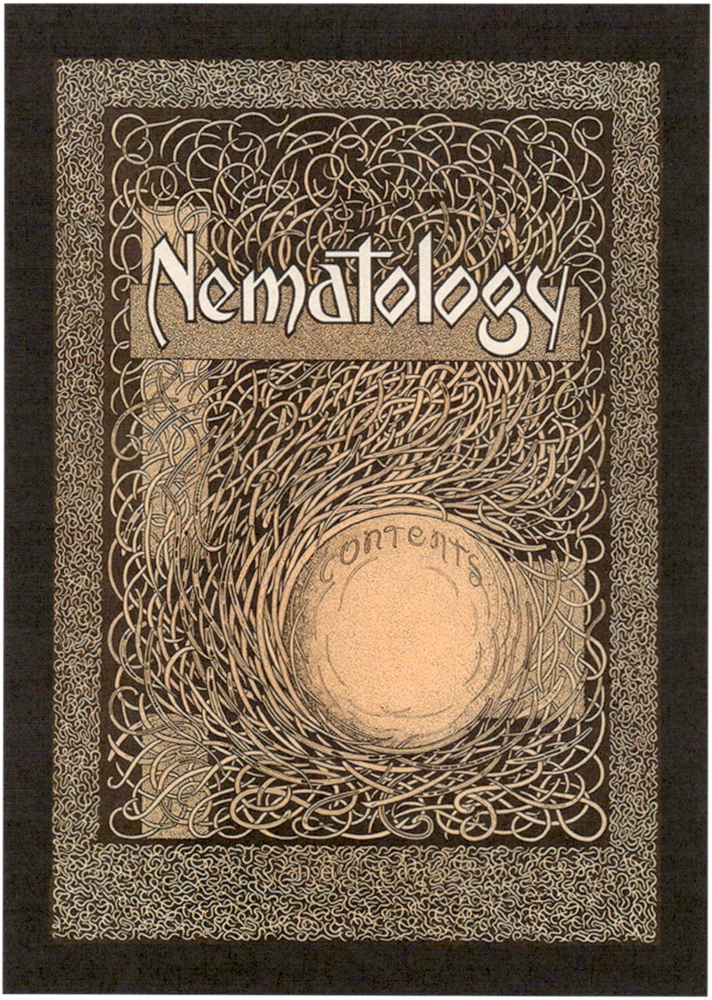
Nematology
CONTENTS

Round cyst nematode Endoparasite

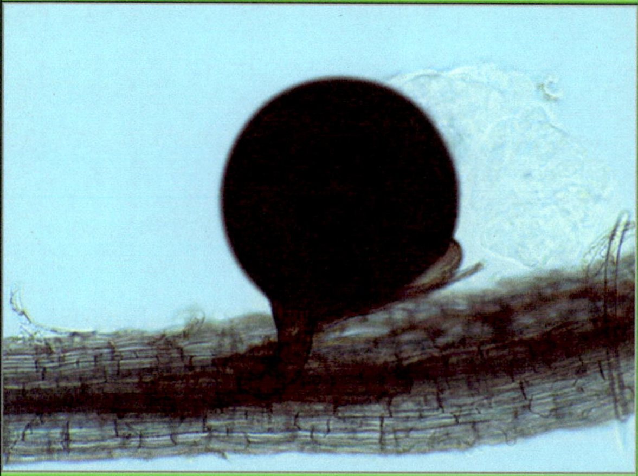

***Globodera* spp.**

Round cyst nematode: mature female pearly white, globose, dead female tans from light to dark brown, several hundred unhatched eggs with second-stage juveniles remain protected within the cyst, two long, convoluted ovaries, unique terminal area with cuticular pattern. Widely distributed, numerous hosts, particularly solanaceous plant, potato, tobacco, tomato, pepper, eggplant and others.

Ulrich Zunke

①

Nematology
CONTENTS

Root-knot nematode Endoparasite

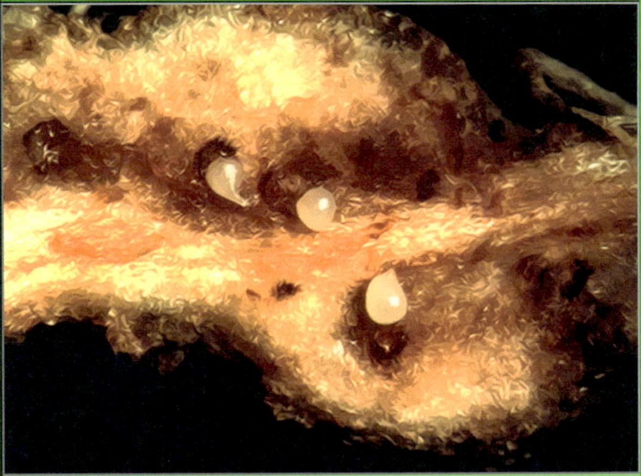

***Meloidogyne* spp.**

Root-knot nematode female: pear-shaped, pearly white, eggs deposited externally in a gelatinous matrix, six large rectal gland cells secrete the matrix through the anal pore; two long, convoluted ovaries, unique fingerprint-like perineal pattern. Worldwide in distribution, numerous hosts from nearly every plant family.

①

Nematology
CONTENTS

Sheath nematode Ectoparasite

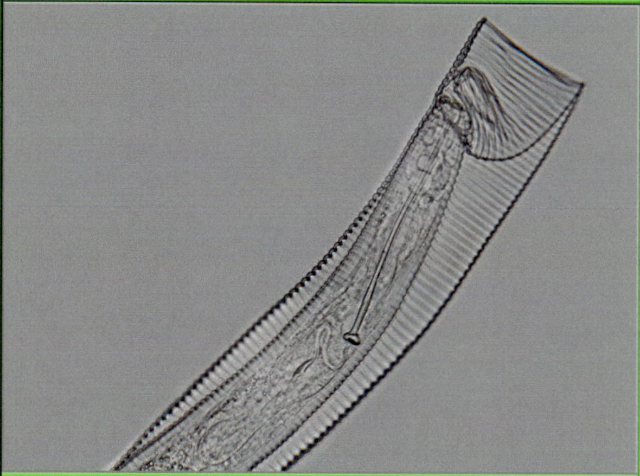

***Hemicycliophora* spp.**

Sheath nematode: moderately long, thick body, heavy annulations, enclosed in a sheath, head and/or tail may be withdrawn into the cuticle, stylet long, strong and slightly curved, procorpus fused with metacorpus, tail of female pointed, males have a small bursa. Widely distributed on maize, grasses, blueberry, cranberry, maple, willow, pine, and numerous woody plants.

①

Nematology
CONTENTS

1620-1671 France

Pierre Borellius

A chemist, botanist, and physician, he was the court doctor for the King of France, Louis XIV. He discovered the vinegar eelworm, Turbatrix aceti, in 1656. Borellius was the first to write about the use of the microscope in medicine.

① 34

Nematology
CONTENTS

1859-1917 Switzerland - Brazil

Émilio August Göldi

Studied in Jena, Germany with Ernst Haekel and accepted an invitation to work at the National Museum of Brazil. While in Brazil he described the genus Meloidogyne and its type species M. exigua, the coffee root-knot nematode. Dr. Göldi also established the connection between root-knot nematodes and vascular wilt fungal diseases.

①

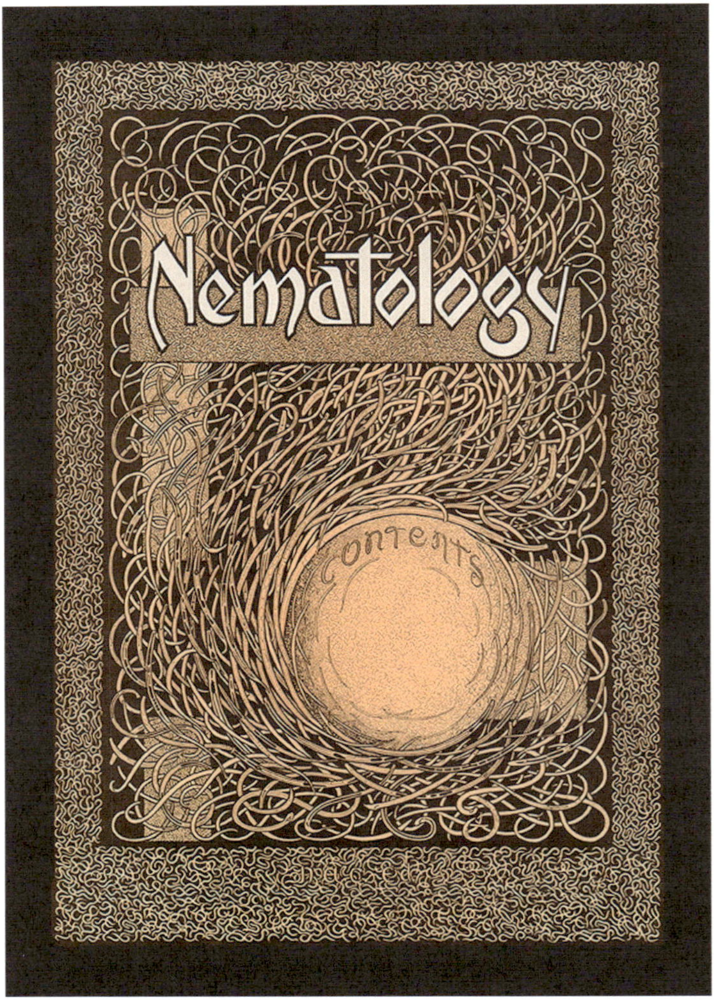
Nematology
CONTENTS

1848-1920 German - Heidelberg

(Johann Adam) Otto Bütschli

He wrote a monograph on the free-living nematodes and noted that these small organisms permitted the obsevation of living cells.

①

Nematology
CONTENTS

1889-1940 Russian

Ivan N. Filipjev

A prominent Russian nematologist and the author of the first general taxonomic classification of nematodes, he published 40 papers on nematology and described more than 150 new species. Prof. Filipjev established the first nematode collection in Russia and was highly regarded. Unfortunately he was executed by the Stalin Regime for his "counter-revolutionary activities."

①

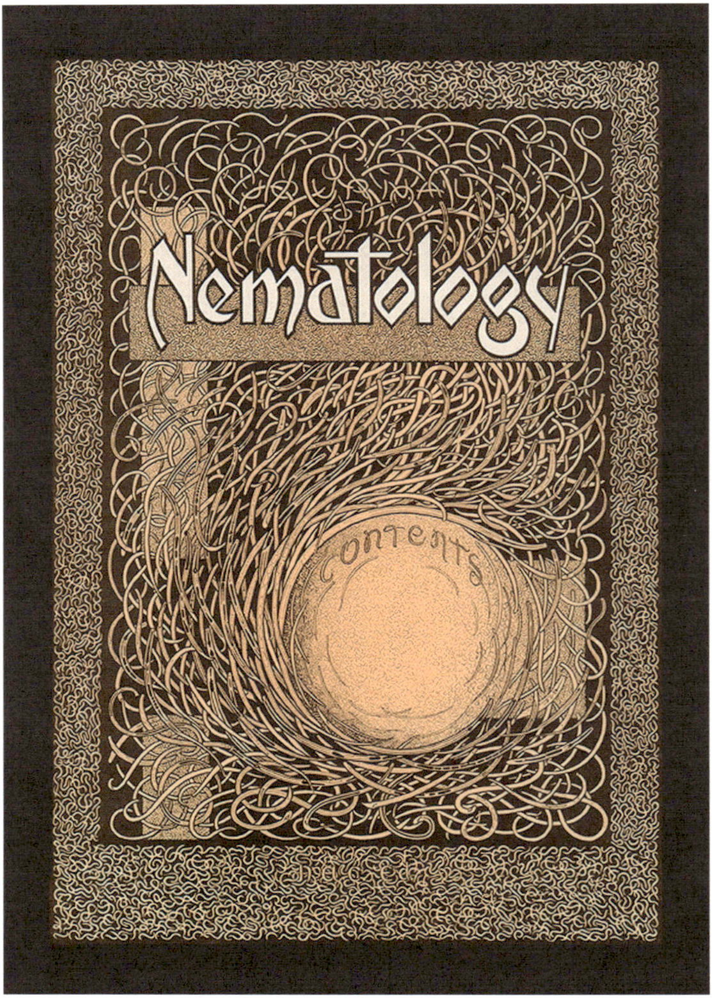
Nematology
CONTENTS

1837-1915 English - MD

Henry Charlton Bastian

A British physiologist and neurologist who had a keen interest in abiogenesis. He believed that he had witnessed the spontaneous generation of life from non-living matter. In the nematological world he is known for his monograph on the Anguillilidae in which he described 100 new species of nematodes, mainly free-living forms.

①

Nematology
Contents

1771-1832 Swedish/German

Karl Asmund Rudolphi

Professor of Anatomy at the University of Greifswald where he proved that plant cells had independent cell walls. His book entitled, "Enterozoorum Sive Vermium Instatinalium Historia Naturalis," was considered to be his greatest contribution to science. He is considered to be the father of helminthology.

①

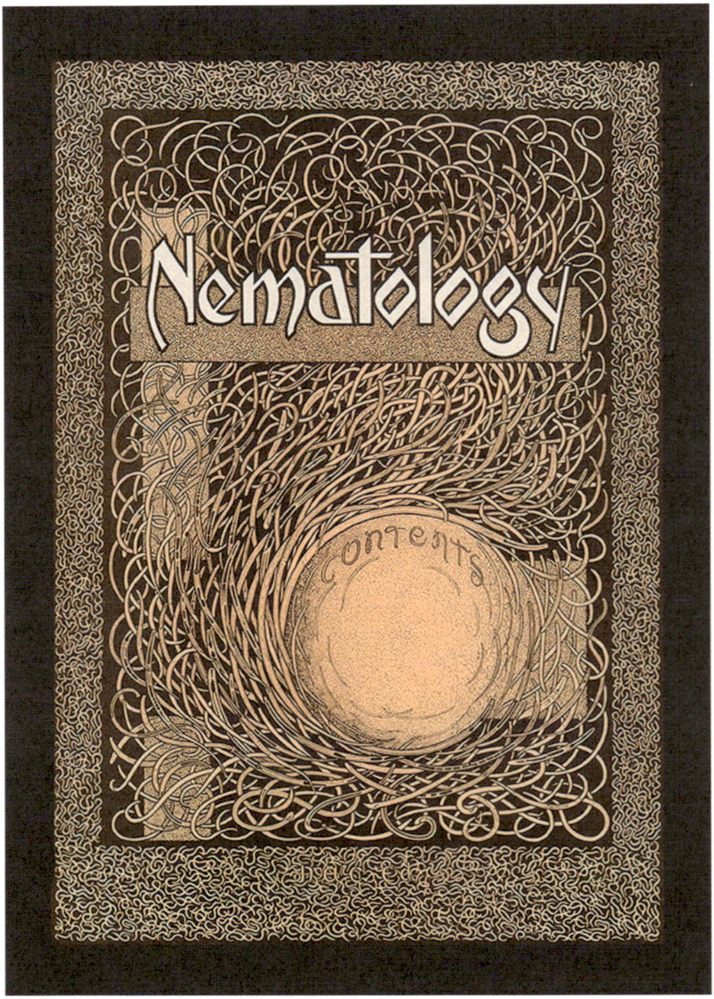
Nematology
CONTENTS

1707-1778 Sweeden - Uppsala

Carl Linneaus

The father of modern taxonomy included in his descriptions of numerous plants and animals that of Ascaris lubricoides, the human intestinal roundworm, and the paste worm, Chaos redivivus (now Panagrellus redivivus). His system of naming organisms with a Latinized binomial of genus and species gave him the role of "type specimen" for the species "Homo sapiens." He was one of the most acclaimed scientists in Europe at the time of his death.

①

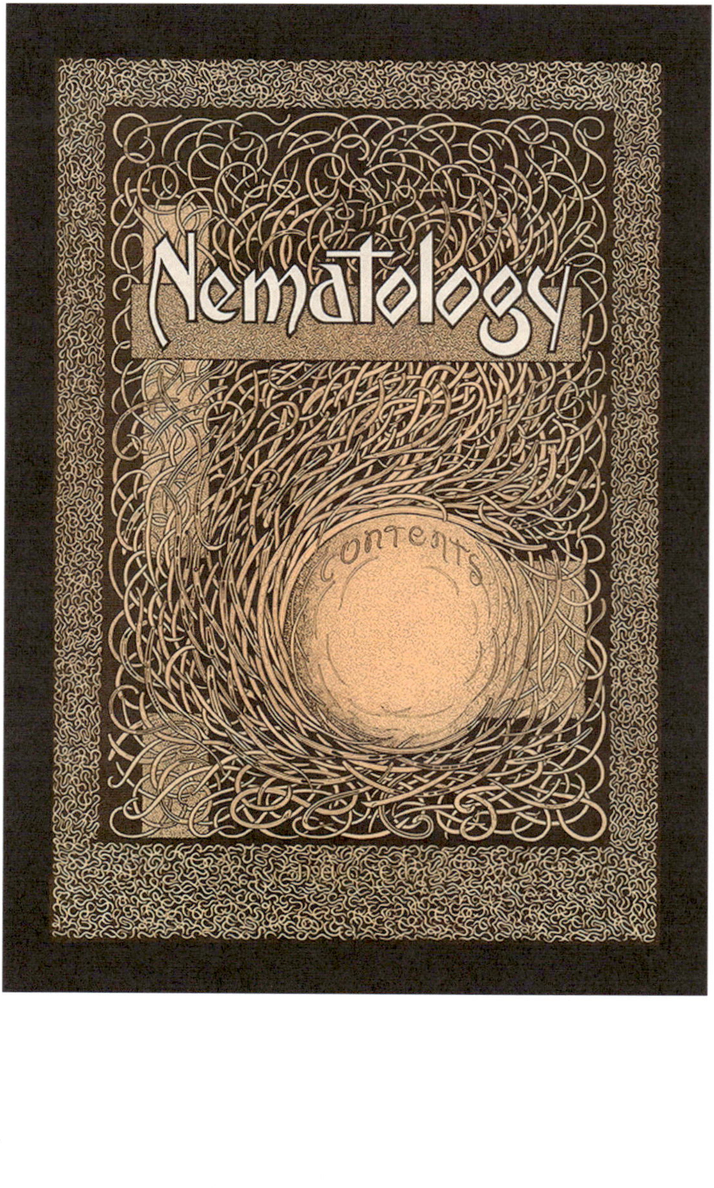
Nematology
CONTENTS

1713-1781 English - The Royal Society

John Turberville Needham

An English ameteur scientist and Catholic priest, he discovered the first plant-parasitic nematode, Anguina tritici on wheat. When he teased apart cockles he found microscopic worms and thought that he had proved Spontaneous Generation of Life. These worms have the ability to become anhydrobiotic and thereby suspend their life until they become rehydrated.

①

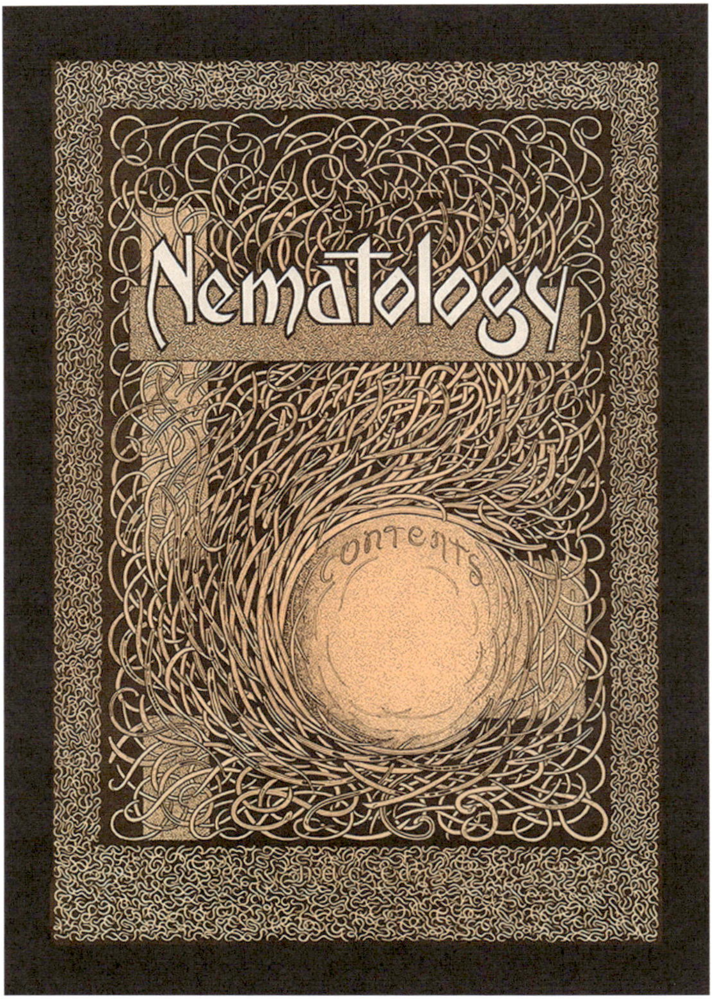
Nematology
CONTENTS

1851-1910 Dutch - Indonesia

Melchlor Treub

A botanist to the Dutch West Indies (Indonesia), born in The Netherlands. His work with tropical flora in the botanical gardens at Bogor helped create an institution that continues to be of world renown. Treub discovered that the Sereh disease on sugarcane was caused by a root-knot nematode that he named Meloidogyne javanica.

①

Nematology
contents

1859-1932 USA – USDA

Nathan A. Cobb

The father of nematology in the U.S. described more than 1,000 new species. He earned his Ph.D. in ten months at he University of Jena. He fathered agricultural research in Australia. Dr. Cobb worked on the diseases of sugarcane in Hawaii and established the Nematology Research Division within the USDA at Beltsville, Md.

①

Nematology
CONTENTS

1921-2005 USA - NCSU

Joseph (Joe) N. Sasser

Inspired by Al Taylor, Joe Sasser earned a Ph.D. under the tutelage of Dr. Benjamin G. Chitwood at the University of Maryland. He gained notoriety for the International Meloidogyne Project which was funded by USAID and was credited for improving the science of Nematology in more than 70 developing countries.

①

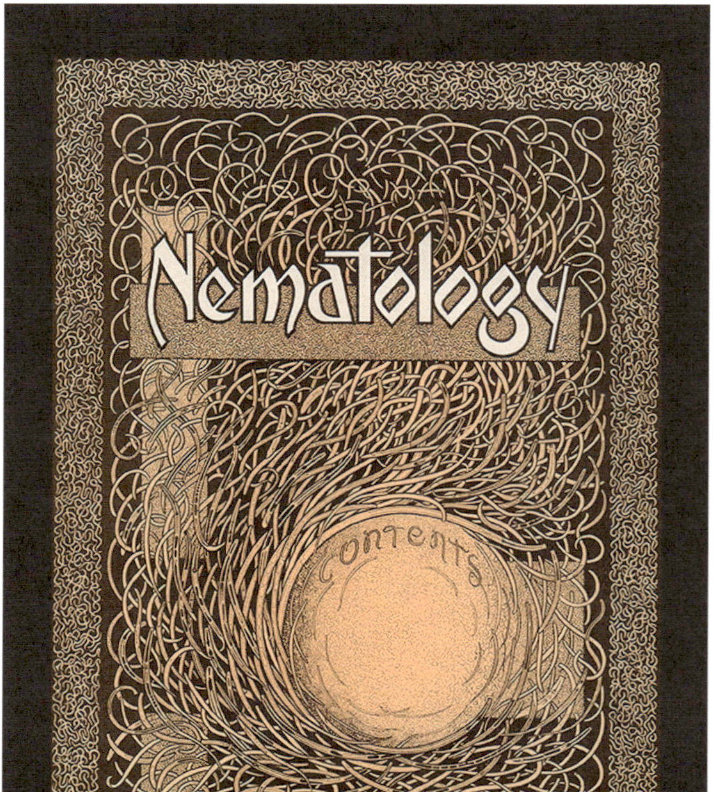
Nematology
CONTENTS

1901-1988 USA - USDA

Albert (Al) L. Taylor

His interest in archery led to a meeting with Gerald Thorne which developed into a distinguished career in nematology. Mr. Taylor established the fourth nematology field station at Tifton, Ga. where he made significant contributions to nematode taxonomy and host interactions. His work on the chemical control of nematodes helped establish the science of nematology as a discipline.

①

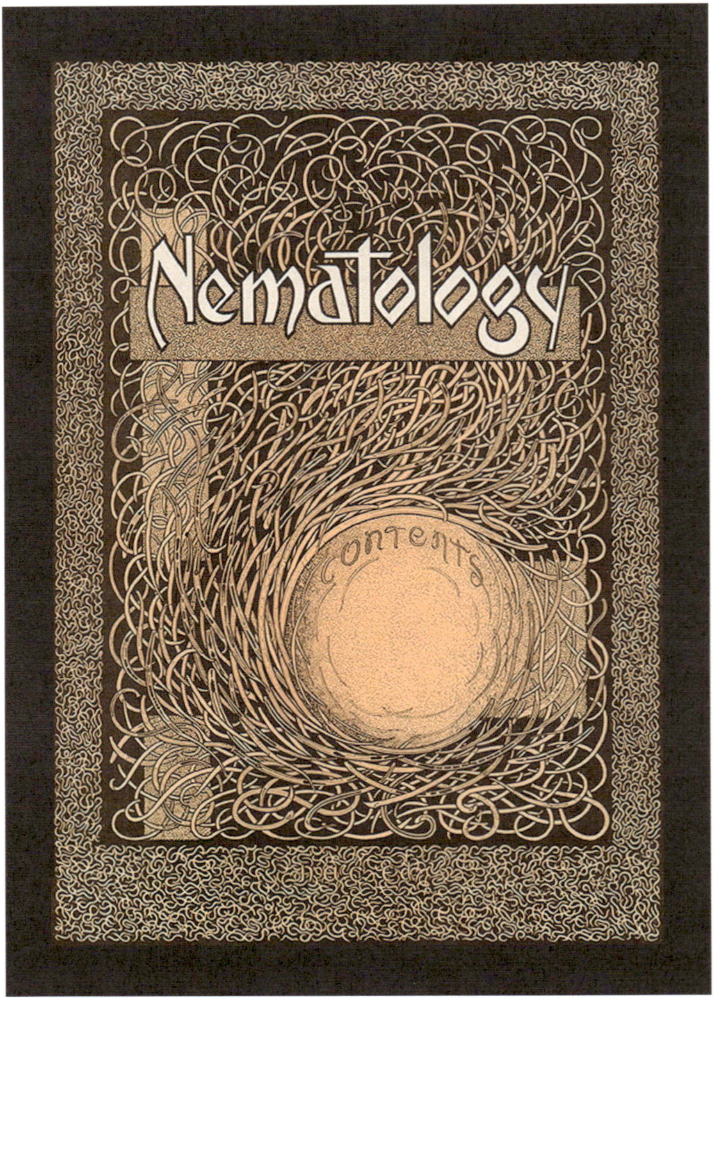
Nematology
contents

1927- USA - NCSU

Hedwig Hirschmann Triantaphyllou

A morphologist and taxonomist who married geneticist and nematologist, Tasso Triantaphyllou, was the fourth female professor at North Carolina State University. Together with her husband and Dr. Joe Sasser she participated in the International Meloidogyne Project that greatly improved nematology in developing countries. She was named a fellow by the SON.

①

Nematology
Contents

1926-2009 USA - NCSU

A. C. (Tasso) Triantaphyllou

Originally from Greece, Dr. Triantaphyllou earned a Ph.D. from North Carolina State University where he became a professor of genetics. Professor Triantaphyllou elucidated the cytology of many nematodes, including the root-knot and cysts nematodes. He teamed up with Dr. Joe Sasser on the International Meloidogyne Project to improve nematology in the developing world.

①

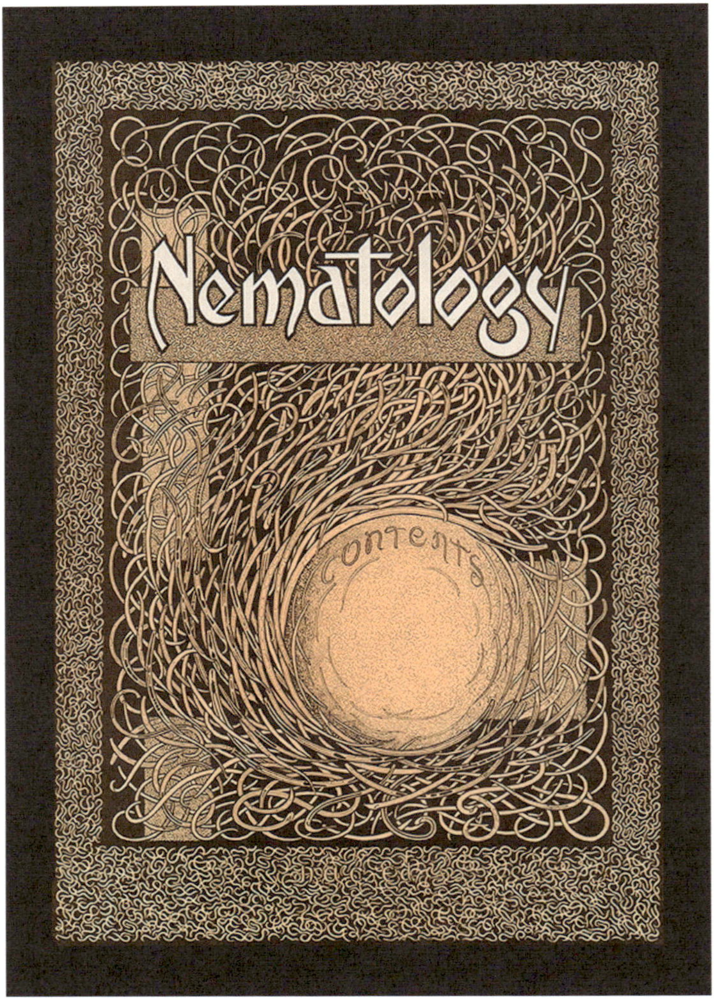
Nematology
CONTENTS

1936- USA

George O. Poinar Jr.

An entomologist who has described numerous nematodes that parasitize insects, he extracted DNA from fossilized insects found in amber and popularized the idea of regenerating animals from their preserved DNA. Poinar is a prolific writer with more than 400 research publications and several books.

Nematology
CONTENTS

1952- USA – VA Tech

Jonathan D. Eisenback

A taxonomist specializing in the morphology of the root-knot nematodes, making him a "Meloidogynecologist," he began his career with the International Meloidogyne Project with Dr. Joe Sasser. He is best known for his SEM of root-knot nematodes, description of several new species, stylet extraction technique, and his high-resolution photomicrographs. Jon served the Society of Nematologists as president and was honored as a Fellow.

①

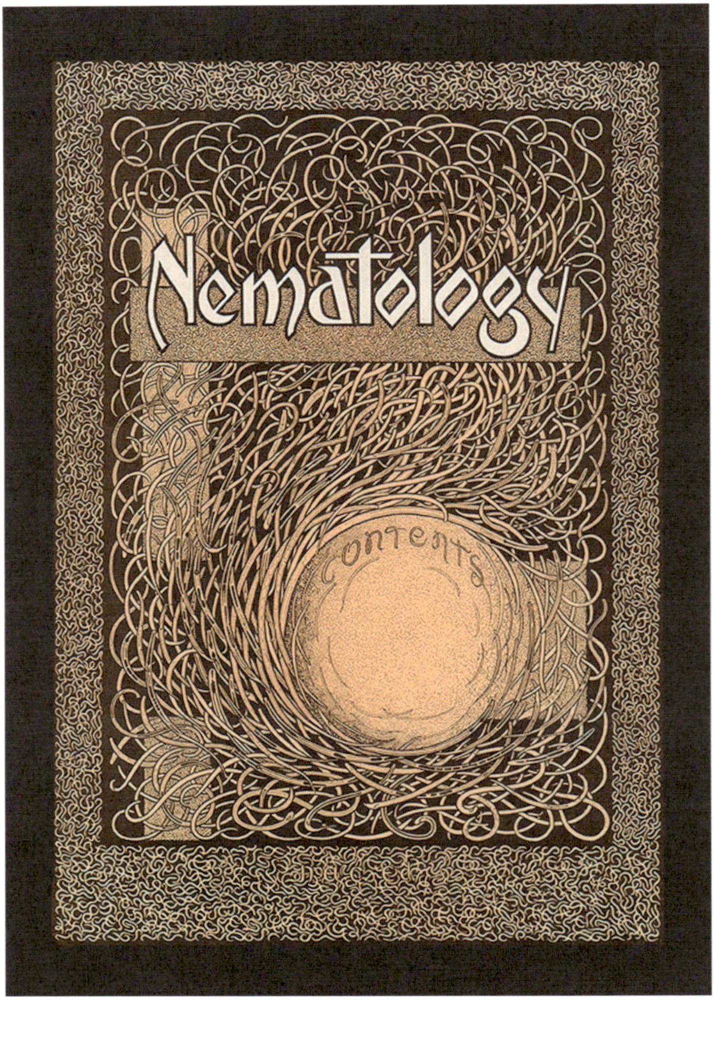
Nematology
CONTENTS

Aphelenchoides spp. Ectoparasite

Angular leaf spots

Foliar nematodes, Aphelenchoides spp., penetrate the open stomates of the leaves of plants and feed on the cells inside the leaf. The damaged cells turn from dark green, to light green, to yellow, to light brown, to dark brown. These spots are often angular because the nematodes cannot cross over the major veins. Ferns, begonias, gloxinia, dahlia, cyclamen, hosta, and many other plants are suceptible to these nematodes.

① 50

Nematology
CONTENTS

Meloidogyne spp. Root system

Galls on roots

Root-knot nematodes usually form characteristic knot-like galls on the roots of infected plants. These galls are sometimes confused with those caused by nodulating bacteria, but usually they wrap around the entire vasular cylinder rather than protruding only from one side of the root.

①

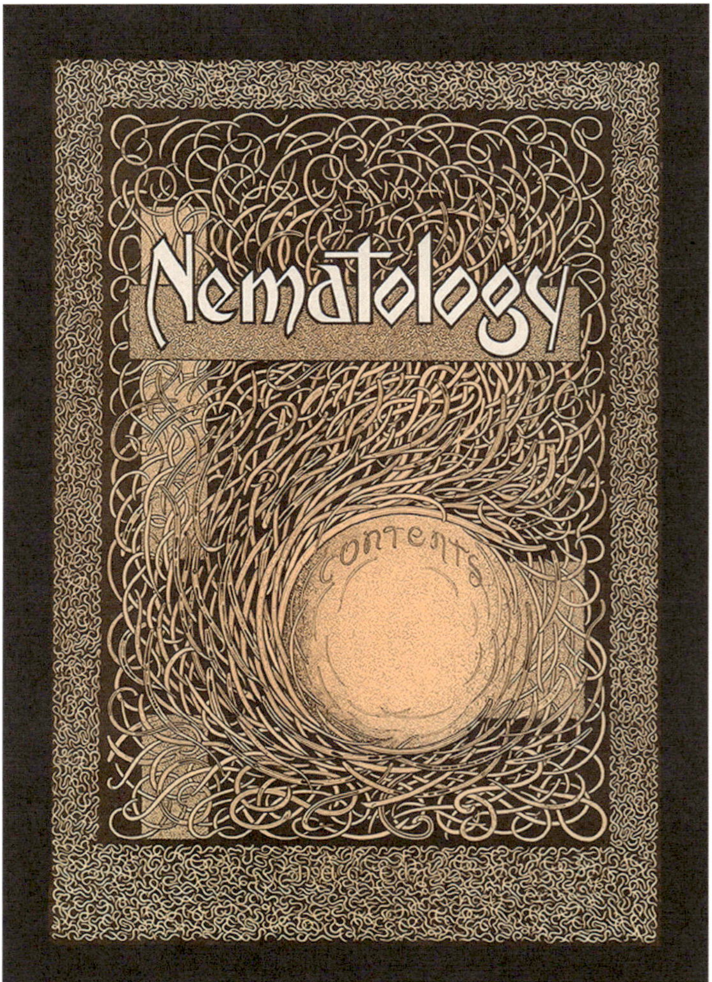
Nematology
Contents

Lesion nematode Root system

Root lesions

Lesions on the roots are symptoms of damage caused by the migratory endoparasitic nematodes like lesion nematodes, *Pratylenchus* spp., and other nematodes such as *Radopholus similis* and *Hirschmanniella* spp. The lesion begins in the epidermal cells as light brown spots that darken with age and severity of the attack.

Society of Nematologists Slide Collection

①

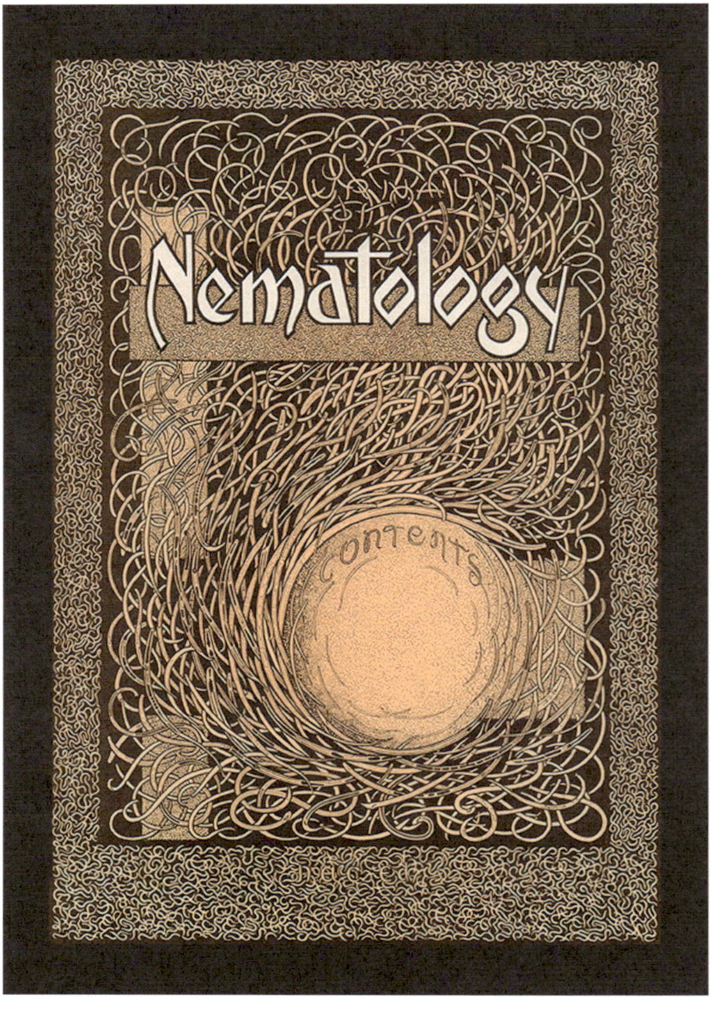
Nematology
CONTENTS

Cuticle

Surface structure

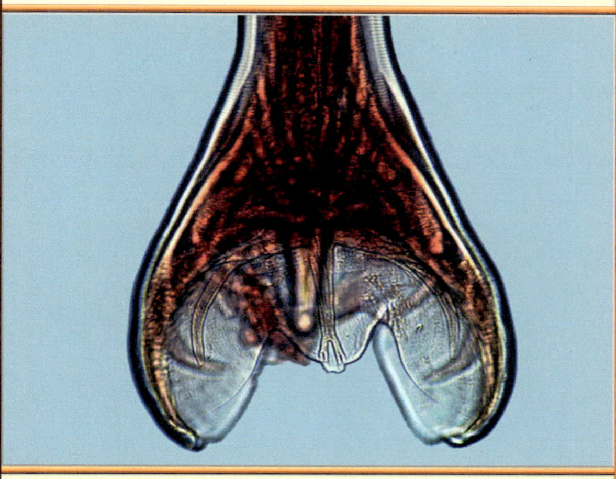

Bursa

Expanded caudal alae in male nematodes are copulatory cuticular structures that are used to grasp the female during copulation. The bursa is formed from two lateral lobes and a third dorsal lobe in some species. The lobes are innervated with finger-like rays that are important taxonomic features.

①

Nematology
CONTENTS

Body wall

Cuticle

Perineal pattern

Cuticular pattern in the perineal area (vulva and anus) of Meloidogyne females composed of fine, wavy, finger-print-like striations that are more or less species characteristic. The shape of the dorsal arch, the occurrence of distinct lateral fields, the presence of punctations in the tail terminus, and the size of the phasmids are useful characters that make-up the pattern.

DECK TWO

1926-1998 Australia - CSIRO

Alan F. Bird

His Ph.D. thesis at the University of Edinburgh on the cuticle of nematodes resulted in papers published in the journals "Nature" and "Science." Dr. Bird spent much of his career in Australia at CSIRO, Divisionof Horticultural Research where he pioneered the electron microscope to study the morphology of nematodes. He contributed to nematode ecology, nematode-microbe interactions, ultrastructure, and behaviour.

② 1

1948-2011 UK - Rothamsted Research

Brian Kerry

Brian was appointed head of the Entomology and Nematology Department at Rothamsted in 1987 and later became the Associate Director of Rothamsted Research. His research on the biological control of nematodes was widely esteemed by his colleagues around the world and for which he was recognized by the Queen of England, the Society of Nematologists and the Organization of Tropical American Nematologists.

②

1933-1998 U.S.A - Univ. Mass.

Burt Zuckerman

Burt pioneered the use of nematodes as models for understanding basic biological processes, a strategy still being used in developmental biology and neuroscience. He was a prolific scientific writer and editor, having been credited with seven books, several laboratory manuals, three patents for novel bio-pesticides, and numerous editorial boards in plant nematology. He was also famous for his "comb-over" hairstyle.

② 3

1707-1778 Sweeden - Uppsala

Carl Linneaus

The father of modern taxonomy included in his descriptions of numerous plants and animals that of Ascaris lubricoides, the human intestinal roundworm, and the paste worm, Chaos redivivus (now Panagrellus redivivus). His system of naming organisms with a Latinized binomial of genus and species gave him the role of "type specimen" for the species "Homo sapiens." He was one of the most acclaimed scientists in Europe at the time of his death.

②

1801-1860 France

Felix Dujardin

A French biologist who was a professor of geology and mineralology at the University of Toulouse, but is remembered for his research with protozoans and invertebrates. His publication entitled, "Historie naturelle des helminthes ou vers intestinaux," contained several descriptions of new species and some very nice drawings of nematodes.

②

1814-1899 Italy

Francesco Miná Palumbo

Both a medical doctor and professional agronomist, he became a naturalist who studied the hydrology, climate, botany, and zoology of Sicily, Italy. He published a report about nematodes on the roots of grapevines and compared the swellings on roots stimulated by nematodes with those caused by Phylloxera, an insect pest related to aphids.

1853-1896 Germany - Univ. Götttingen

Georg Liebscher

Professor and Director of the Agricultural Research Institute at Göttingen, Germany where he described the pea cyst nematode, Heteredera goetingianna. His dissertation was titled, "On the relations of Heterodera schachtii for beet fatique." He was best known for his studies on the fertilzer needs of crop plants.

②

1944-2012 New Zealand - Soil Bureau

Gregor William Yeates

Soil zoologist and ecologist who was "considered the world's leading authority in soil nematode ecology" worked at the Soil Bureau, a division of DSIR, where he published over 200 papers and described over 100 species in Longidorus, Xiphinema, Hemicycliophora, Trischistoma, Dorylaimida and Iotonchus. Yeates's early work in sand dunes was the first detailed work on nematodes in non-agricultural settings.

②

1950- USA – Texas A & M

James (Jim) L. Starr

His research on the identification, characterization, and deployment of resistance to nematodes in important crops included root-knot nematodes (RKN) on peanut, RKN and reniform on cotton, RKN on pecan and forage legumes. He also investigated the relationship between the polyploidy of giant cells and nematode development; and investigated why two-thirds of the nearly 100-fold increase in gene copy number in giant-cell is not required for development.

②

1943- UK – Rothamsted Research

Kenneth (Ken) Evans

Ken's expertise in the laboratory and field on cyst nematodes was fundamental in understanding the importance of these nematodes. His research on hatching factors, resistance and tolerance, biological control agents, population dyanmics, fungus-nematode interations, trap crops, and prescription application of nematicides were all aimed at reducing the economic damage of nematodes. Together with his passion for nematodes, Ken is a keen golf player.

NEMATO-lab, Coimbra, Portugal

② 10

1948-2012 India

M. R. Siddiqi

He was a prolific taxonomist and morphologist with 284 research papers dating from 1959, detailing a proposal for one new subclass, 15 new orders, eight new superfamilies, 31 new families, 31 new subfamilies, 248 new genera, and 671 new species and subspecies of Nematoda. His magnum opus was "Tylenchida Parasites of Plants and Insects," acclaimed as a "Milestone in the history of Nematology."

②

1943-

Portugal – Oeiras

Maria Antónia Bravo

Senior Nematologist Researcher at the National Agr. Research Station (now INIAV). She worked mainly on the taxonomy of plant-parasitic nematodes that vector viruses (Longidoridae). She elucidated their distribution and described several new species. She was also involved on the first detection of the pinewood nematode (Bursaphelenchus xylophilus) in Portugal, and in Europe, in 1999.

Paulo Vieira and Manuel M. Mota

② 12

1927-2010 France - ORSTROM

Michel Luc

Michel Luc was an early plant nematologist who worked with ORSTROM in France, establishing the first laboratory in the research station of Adiopodoumé, near Abidjan, Ivory Coast and another lab at Dakar, Senegal. He was the founding editor of Revue de Nématologie and was a widely recognized taxonomic authority on several groups of plant-parasitic nematodes.

②

1950- USA – Rutgers Univ.

Randy Gaugler

A distinguished professor and a fellow of four scientific associations, including SON; he is a foremost authority on insect-parasitic nematodes and he has made significant contributions to virtually every facet of their biology, especially their behavioral ecology. He has edited three books on nematode behavior and entomopathogenic nematology. His studies have been cited more than 15,000 times.

Isik Unlu

② 14

1665-1703 England - The Royal Society

Robert Hooke

In 1667, he discovered the paste eelworm or beer mat nematode, now known as Panagrellus revidivus. It was first decribed by Linnaeus one hundred years later under the Latinized binomial scientific name of Chaos redivivum. Aquarium enthusiasts use it to feed their small fry and refer to it as "micro-worm." Females deliver live young rather than laying eggs.

1946- UK – Rothamsted Research

Roland "Rolo" N. Perry

His research interests are on plant-parasitic nematodes, especially hatching, sensory perception, behaviour, and survival physiology. He authored or co-authored numerous books, over 40 chapters and reviews, and over 100 research papers. Rolo was EIC of Nematology, the Russian J. of Nematology, and Nematology Monographs and Perspectives. He is a Fellow of the SON, the ESN, and Honorary Member of the Russian Society.

②

1951- USA – NMSU

Stephen H. Thomas

Distinquished Service Professor of Nematology: his research interests include host-parasite relationships involving plant-parasitic nematodes, soil nematode ecology, nematode interactions with weeds and fungal pathogens, cropping systems for sustainable plant-pathogenic nematode management, and nematodes as indicators of soil health. He served as President of the Society of Nematologists during 2014-15.

②

1935- Portugal - Univ. Coimbra

Maria Susana Santos

Pioneer Portuguese nematologist who made a remarkable contribution to the knowledge of the taxonomy, biology and ecology of plant-parasitic nematodes. She described several species of Meloidogyne, Xiphinema and Trichodorus and was the first to report Globodera pallida in Portugal. As a Professor at Coimbra University, she mentored several generations of aspiring nematologists.

NEMATO-lab, Coimbra, Portugal

②

1950- Belgium – Ghent Univ.

Wilfrida Decraemer

She served as Director and Senior Full Professor in the International Master of Science in Agro- and Environmental Nematology at Ghent University. She is an international expert on nematode taxonomy and has made major contributions to many groups, including the Trichodorids, the Longidorids, entomopathogenic nematodes, and the free-living marine nematodes.

② 19

1943- USA – UC Riverside/Davis

Howard Ferris

His research intersests include soil ecology; decomposition and nutrient cycling; contribution of nematodes tothe structure and function of the soil food web; population biology and population assessment of soil nematodes; and integrated pest management. Born in England, raised in South Africa and Zimbabwe, schooled in New Zealand, the Netherlands, Zimbabwa, and North Carolina, he worked at U.C. Riverside and U.C. Davis.

② 20

1932- USA - N.C.S.U.

Kenneth R. Barker

He utilized hundreds of microplots to study the population dynamics of plant-parasitic nematodes, and developed the semi-automatic elutriator for the extraction of nematodes from soil. Instrumental in the organization of the International Federation of Nematological Societies, he served as its first president. He was also Editor-in-Chief of the Journal of Nematology, President of the Society, Fellow of SON, and named as an Honorary Member.

1927-2017 USA – Perdue Univ.

Virginia Ferris

She was the first woman appointed to the Purdue Agriculture Faculty, paving the way for women in agriculture. Her interest in nematodes came as an assist. professor at Cornell and continued to increase throughout her career. Virginia served as President of the Society of Nematologists and was honored as a Fellow and Honorary Member. She often asked the most thought-provoking question at sessions of the SON, right up until her last years.

②

Fungus Egg parasite

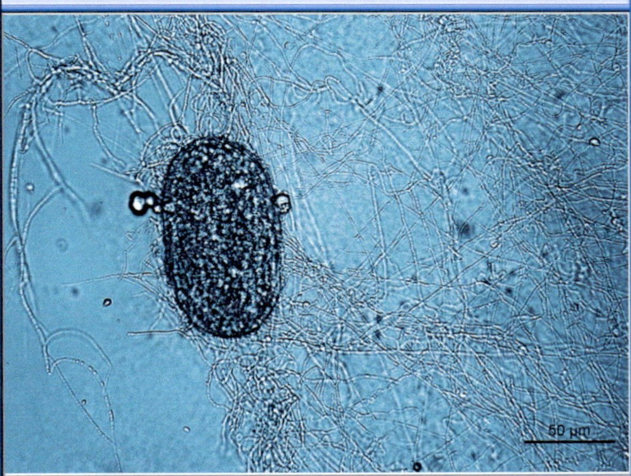

Pochonia chlamydosporia

Endophytic ubiquitous facultative nematophagous fungus, parasitic on nematode eggs of the genera Globodera, Heterodera, Meloidogyne, Nacobbus and Rotylenchulus. It is one of the most studied biological control agents against plant (semi-) endo-parasitic nematodes and has been developed commercially in Brazil, China, Cuba and Italy.

NEMATO-lab, Coimbra, Portugal

② 23

Chemical Seed treatment

Tioxazafen

A seed treatment nematicide from Monsanto for controlling nematodes on corn, soybean, and cotton. The mode of action through the disruption of the mitochondria of nematodes is unique for this product. It is effective against root-knot, soybean cyst, lesion, and reniform nematodes. The trade name is a part of the Acelleron seed applied solutions called "Nemastrike."

② 24

Chemical Non-fumigant

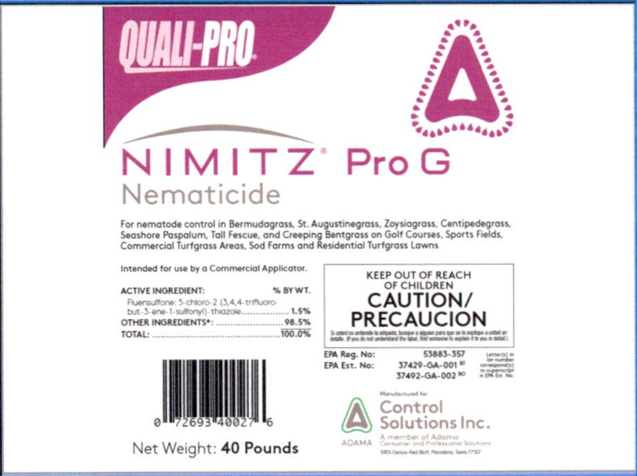

Fluensulfone

A nematicide from ADAMA for controlling nematodes in agronomic crops and turfgrasses. It is effective against root-knot, sting, and lesion nematodes on cucumber, melons, squash, tomatoes, okra, eggplant and peppers. Applications are broadcast followed by irrigation to get the product into the root zone and soil. Effectiveness may be reduced in muck or high clay content soils. Applied at a rate of 60-120 pounds of product per acre.

②

Biological Non-fumigant

Specimen Label

ZELTO™

Active Ingredient: Heat-killed *Burkholderia spp.* strain A396 cells and spent fermentation media* 94.46%
Other ingredients: 5.54%
Total: 100.00%
*Contains not less than 1,500 Beet Armyworm Killing Units (BAWKU)/mg. Note: The percent active ingredient does not indicate product performance and potency measurements are not federally standardized.

KEEP OUT OF REACH OF CHILDREN
CAUTION

FIRST AID	
If in eyes	• Hold eye open and rinse slowly and gently with water for 15–20 minutes. • Remove contact lenses, if present, after the first 5 minutes, then continue rinsing eye. • Call a poison control center or doctor for treatment advice.
HOT LINE NUMBER Have the product container or label with you when calling a poison control center or doctor, or going for treatment. You may also contact 1-800-222-1222 for emergency medical treatment information.	

EPA Reg. No.: 84059-14

Zelto™ is a trademark of Marrone Bio Innovations, Inc.
Marrone Bio Innovations name and logo are registered trademarks of Marrone Bio Innovations, Inc.

Burholderia rinojensis strain A396

A nematicide from Marrone Bio Innovations Inc. made from heat-killed cells and spent fermentation media of Burkholderia rinojensis strain A396 in a liquid water dispersable formulation that has a stable shelf-life of at least 2 years and a broad tank-mix compatibility with pesticides and fertilizers. It is effective against most plant-parasitic nematodes including root-knot, sting, ring, lance, and stubby root.

Chemical non-fumigant

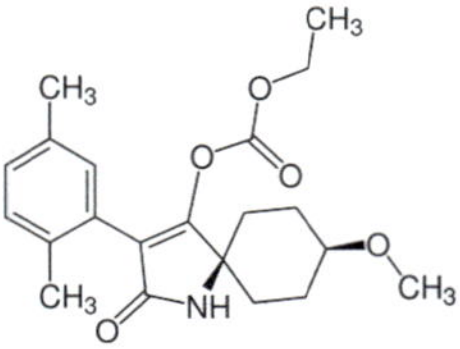

Spirotetamat

Spirotetamat is a lipid biosynthesis inhibitor that supresses lipids resulting in poisoning and subsequent arrest of nematode development folllowing ingestion by feeding on the plant. It moves from leaf to leaf and leaf to root to protect the entire plant where it is effective at suppressing several kinds, including root-knot, ring, lesion, dagger, cyst, stubby root, and citrus nematodes.

② 27

Chemical — non-fumigant

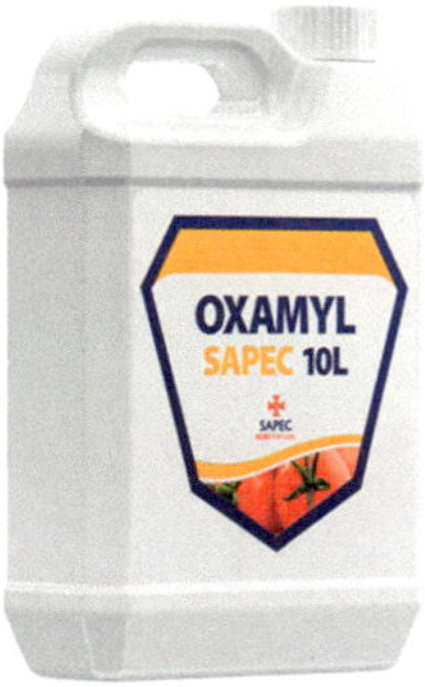

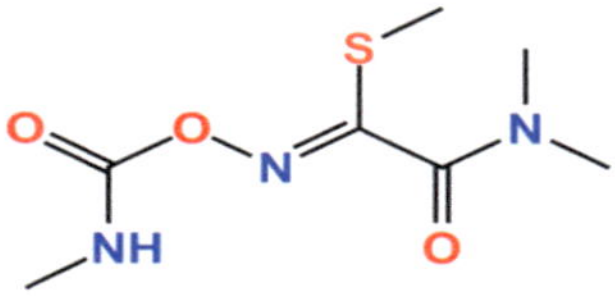

Oxamyl

Oxamyl is a carbamate nematicide that effectively inhibits acetylcholinesterase, acting as a nerve poison. It is commonly sold under the trade name of Vydate and is used for managing plant-parasitic nematodes on a variety of vegetables and fruit crops. Oyamyl can be applied as a foliar spray at planting where it translocates downward to protect the roots or via chemigation where it moves upward to protect the canopy.

②

Fumigant metam sodium

Sodium methyldithiocarbamate

Metam sodium, as commonly named, is a powerful fumigant that is widely used in crop production, especially on potato in Idaho, Washington, Oregon and California to manage plant-parasitic nematodes. It is also often used in the culture of tomato, strawberry, carrot, melon, peppers, grape, citrus, lettuce, asparagus, spinach, tree seedlings, and ornamentals for the same purpose. It is sold under the trade names of Vapam, Metam CLR and Sectagon.

② 29

Chemical non-fumigant

MULTIGUARD PROTECT®

For terrestrial (outdoor) non-food use on established turf on golf course tees and greens, practice greens, spot treatment of fairways, roughs and turf/sod farms; when used as directed, MULTIGUARD PROTECT® controls root infesting plant parasitic nematodes, and suppresses fungal plant diseases such as species of Pythium, Phytophthora, Fusarium and Rhizoctonia.

Active Ingredient	(by weight)
Furfural	90.0%
Other Ingredients	10.0%
TOTAL	100.0%

1 gallon of MULTIGUARD PROTECT® contains 8.68 lbs furfural
1 gallon of MULTIGUARD PROTECT® weighs 9.65 lbs at 68°F

KEEP OUT OF REACH OF CHILDREN

WARNING – AVISO

Si Usted no entiende la etiqueta, busque a alguien para que se la explique a Usted en detalle. (If you do not understand the label, find someone to explain it to you in detail.)

	FIRST AID
If Swallowed	• Call a poison control center or doctor immediately for treatment advice. • Have person sip a glass of water if able to swallow • Do not induce vomiting unless told to do so by the poison control center or doctor. • Do not give anything by mouth to an unconscious person.
If in Eyes	• Hold eye open and rinse slowly and gently with water for 15-20 minutes. • Remove contact lenses, if present, after the first 5 minutes, then continue rinsing eye. • Call a poison control center or doctor for treatment advice.
If on Skin or Clothing	• Take off contaminated clothing. • Rinse skin immediately with plenty of water for 15-20 minutes. • Call a poison control center or doctor for treatment advice.
If Inhaled	• Move person to fresh air. • If person is not breathing, call 911 or ambulance, then give artificial respiration, preferably by mouth-to-mouth, if possible. • Call a poison control center or doctor for further treatment advice

HOTLINE NUMBER: CHEMTREC 1-800-424-9300 (24 hours)

Have the product container or label with you when calling a poison control center or doctor, or going for treatment. You may also contact 1-800-222-1222 for emergency medical treatment information.

NOTE TO PHYSICIAN: Probable mucosal damage may contraindicate the use of gastric lavage.

SEE SIDE PANEL FOR ADDITIONAL PRECAUTIONARY STATEMENTS

Manufactured for Agriguard Company, LLC
Centennial Plaza, Suite 100, 186 North Avenue East, Cranford, NJ 07016

EPA Establishment No. 34704-MS-02 **EPA Registration No. 75753-1**

Net Contents: 2½ Gallons

31694

Furfural

A contact nematicide for established turf on golf course tees and greens, spot treatment of fairways and roughs, and sod farms for the management of plant-parasitic nematodes, furfural's mode of action is by disentigration of the cuticle and subsequent death of the nematode through contact. It is sold under the trade name of Multiguard Protect and has also been shown to suppress several fungal plant pathogens.

Ecosystems Temperate and tropical

Agricultural soils

Soil is the foundation of agriculture and contains mineral components, organic matter, water, and air that allow water retention and drainage, nutients availability, oxygenation, and physical support for the plants. Nematodes are often the most common meiofauna present in agricultural soils. Nematodes from 10 acres of farmland lying head end to tail end would encircle the earth at the equator.

Turfgrass Golf greens and tees

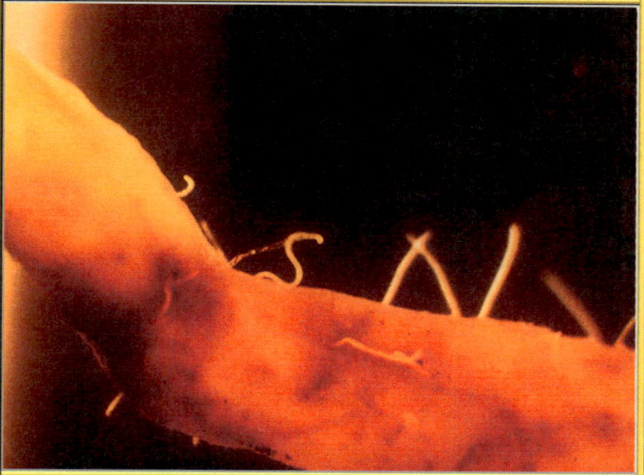

Turfgrass spp.

Golf greens built according to PGA standards are keen habitats for plant-parasitic nematodes and several genera can cause poor root growth. These nematodes cause the turf to become chlorotic, wilt during the hot part of the day, and to eventually thin out and die. The most important nematodes on turf include the ring, sting, lance, stubby root, lesion, and root-knot.

②

Anguina spp. Temperate and Subtropical

Seed heads

Anguina agrostis parasitizes the seed heads of several grass species including Agrostis spp., Bromus spp., Dactylis spp., Festuca spp., Lolium spp., Phleum spp., Poa spp., and Trisetum spp. Symptoms include abnormal leaves, lesions on glumes, elongated glumes, dicolored seeds, and seed galls. Parasitized plants have been described as a new spccies because the glumes were so distorted and elongated.

Aerial Plant leaves

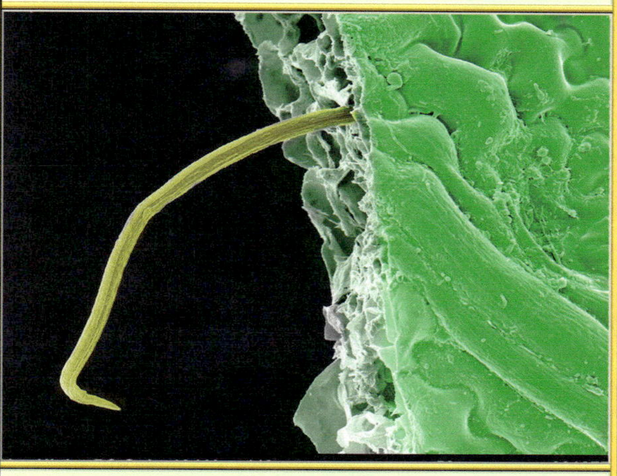

Inside plant leaf

Nematodes are splashed from the soil onto leaf tissue where they enter the wet leaf through stomates and feed within the spongy mesophyll layer of cells. The cells respond by changing colors, first a pale green, then a chlorotic yellow, and then from light brown to dark brown. These lesions are angular in appearance because the nematodes are restricted in their movement by the large viens in the leaf.

②

Heartworm

Dogs and cats

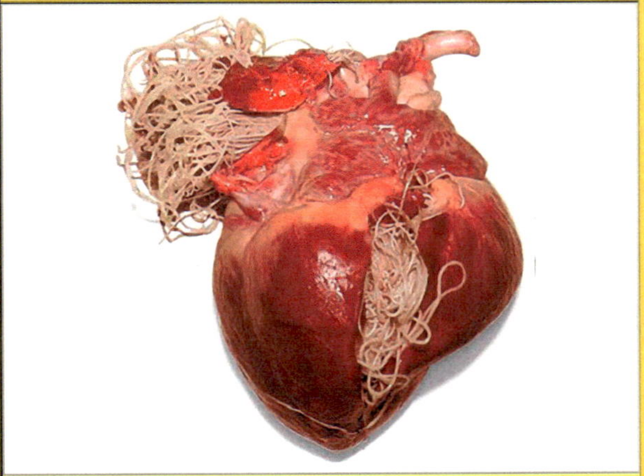

Pulmonary arterial system

Diofilaria immitis lives in the heart and the pulmonary arterial system of dogs, the definitive host; however it can also parasitize cats, coyotes, jackals, foxes, bears, ferrets, seals, sea lions, beavers and in very rare circumstances, humans. Mosquitos are secondary hosts and also serve as vectors. This worm often interacts with an intracellular bacterium, Wolbachia pipientis, as an endosymbiont which makes the disease even more pathogenic.

②

Marine Ice free coastlines

Sandy beaches

Composed of quartz and/or carbonite sands, feldspar, and heavy minerals, sandy beaches have varying grain size that affects porosity, percolation, drainage, and oxygenation. However no matter what these parameters are, nematodes are extremely numerous on sandy beaches, but their role in the ecosystem remains largely unknown, even though they are the dominant meiofauna group in these ecosystems.

②

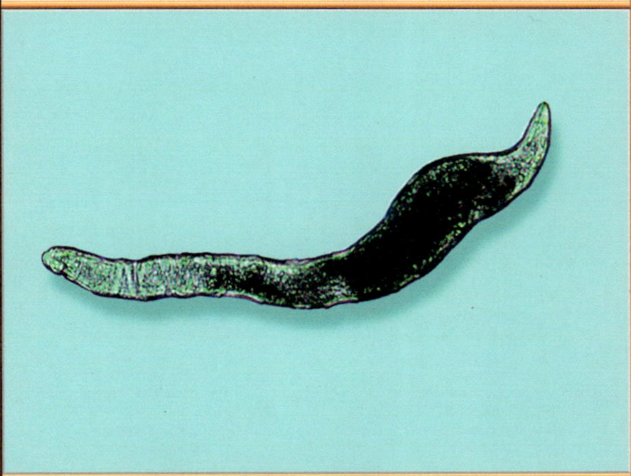

Female intersex

Root-knot females may express secondary sex characteristics of males with a vermiform posterior containing rudiments of spicules and a gubernaculum, (very common in males of Meloidogyne hapla). Sex determination in the root-knot nematodes can be affected by environmental stresses, especially the status of the host plant. As the host deteriorates, a higher percentage of the developing juveniles become males.

Reproductive system

Male

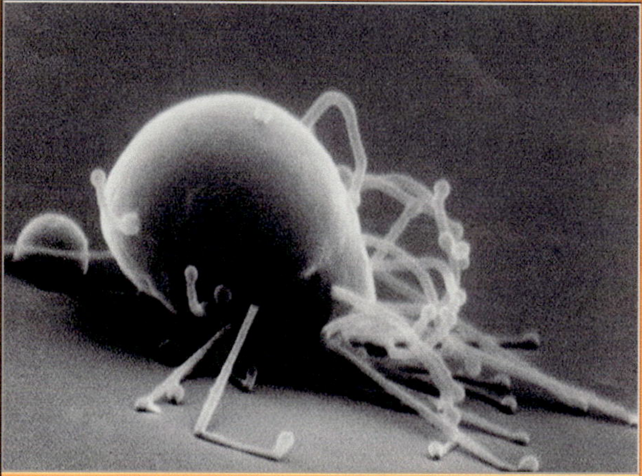

Cyst nematode sperm

Sperm from the spermatheca of the female is composed of a large, round, smooth cytoplasmic region that formed many pseudopodia that enabled it to crawl up the uterus into the spermatheca. After arriving they are no longer needed, and the opposite end forms numerous filopodia that are probably used to attach to the inner walls of the spermatheca.

②

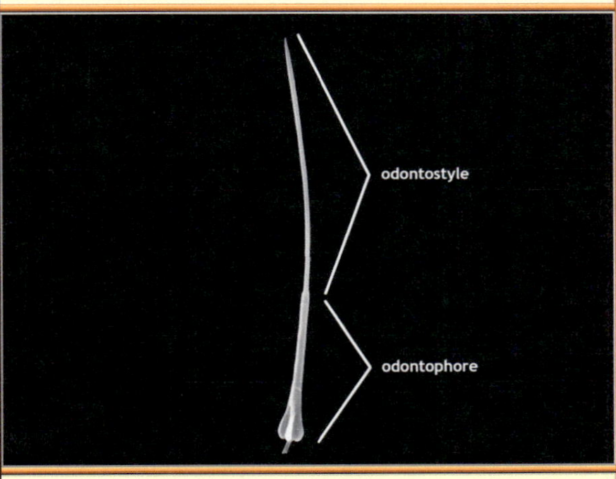

Odontostyle

A hollow, protrusible, needle-like tooth that is used for feeding, with a dorsally oblique aperture derived from a single cell in the anteroir part of the esophagus. Often with a posterior attachement called an "odontophore" or "odontostyle extension."

syn. = odontostylet, spear

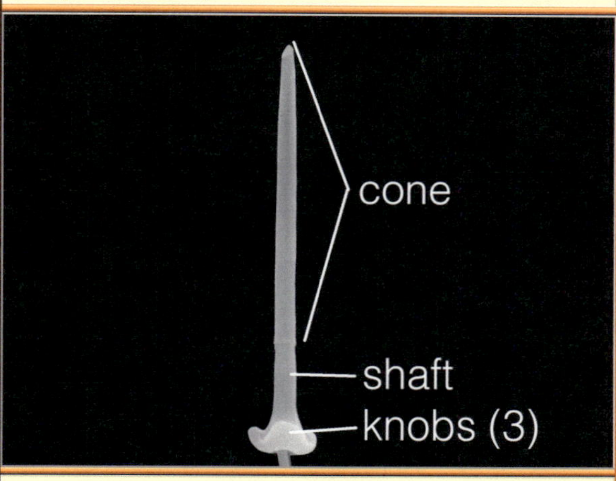

Stomatostyle

A hollow, protrusible, needle-like tooth that is used for feeding through a subterminal, ventral opening; found in plant, fungus, and insect feeding nematodes of the sub-orders Tylenchina and Aphelenchina. It is made-up of three distinct parts: cone, shaft and three knobs - one dorsal and two sub-ventral.

syn. = stomatostyle, stylet, spear

Reproductive system — Female gonad

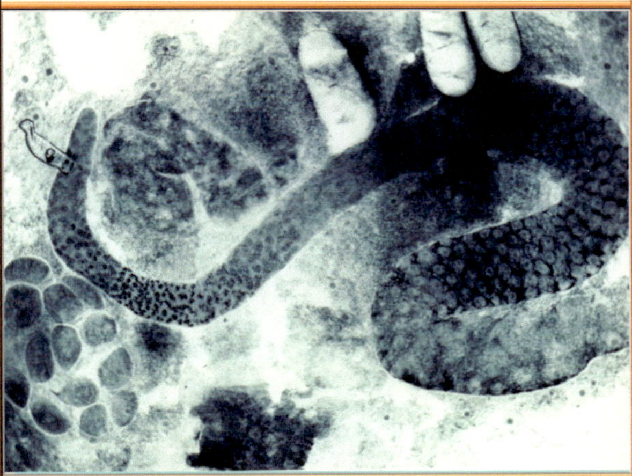

Root-knot female ovary

Approximately 60% of the female gonad is made up of the ovary proper. The anterior end of the ovary contains the cap cell that forms the germinal zone of the ovary and is the source for all of the oogonia. The oogonia are arranged around a rachis in the germinal zone of the ovary where they increase in size to form the oocytes. The remaining posterior portion of the gonad is made up of the spermatheca and uterus.

Cuticle Survival mechanism

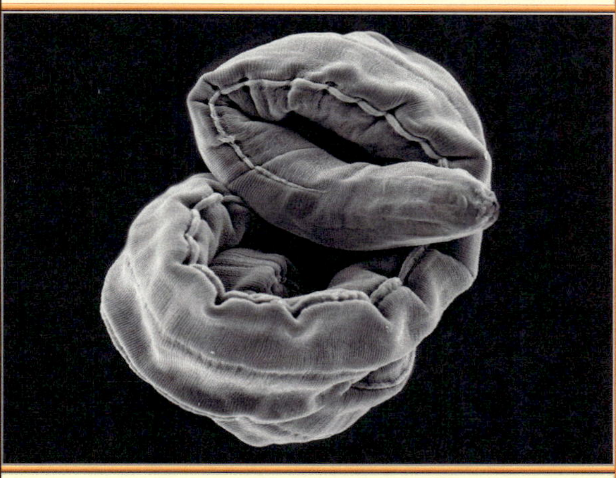

Anhydrobiosis

First described by Antony van Leeuwenhoek more than 300 years ago from nematodes observed in the dried sediment of the gutters on his house, anhydrobiosis means "life without water." Many nematodes are extreme dessication-tolerant and may survive for many years without water. Upon drying, anhydrobiotic nematodes usually assume a compact, spiral shape, but they quickly absorb water within a few minutes and become active again.

② 42

Reproductive system Female gonad

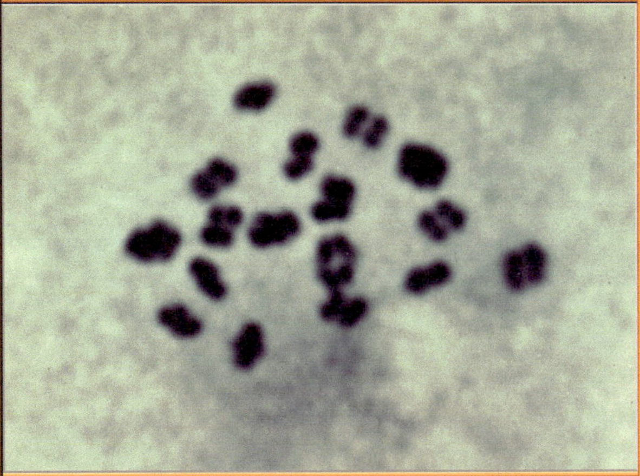

Root-knot tetrad chromosomes

A tetrad, also known as a bivalent chromosome, is the association of two homologous chromosomes that are physically attached to each other allowing for alignment and segregation of the genetic material during the first meiotic division. In the root-knot nematodes, this pairing of homologous chromosome indicates that the species reproduces either by amphimixis or meiotic parthenogenesis.

②

Reproductive system — Male intersex

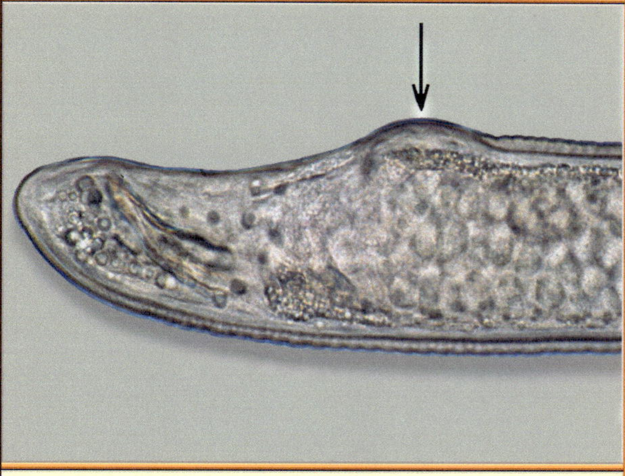

Rudimentary vulva

Root-knot males may express secondary female sex characteristics of females of rudiments of a vulva and vagina, (very common in males of Meloidogyne javanica).

②

Cuticle

Surface structure

Perineal pattern

Cuticular pattern in the perineal area (vulva and anus) of Meloidogyne females composed of fine, wavy, fingerprint-like striations that are more or less species characteristic. The shape of the dorsal arch, the occurrence of distinct lateral fields, the presence of punctations in the tail terminus, and the size of the phasmids are useful characters that make-up the pattern.

②

Stem and bulb Migratory endoparasite

Ditylenchus dipsaci

Primarily a parasite of onion and garlic, but also infects alfalfa and other plants. Females and males are common since reproduction occurs by amphimixis. Approxomately 250 eggs are produced by each female which take about 20 days to develop into second-stage juveniles. These forms either reinfect the host or they overwinter as anhydrobiotic "nema wool" inside dry plant tissue, including seed pods.

②

Wilsonema sp. Free-living

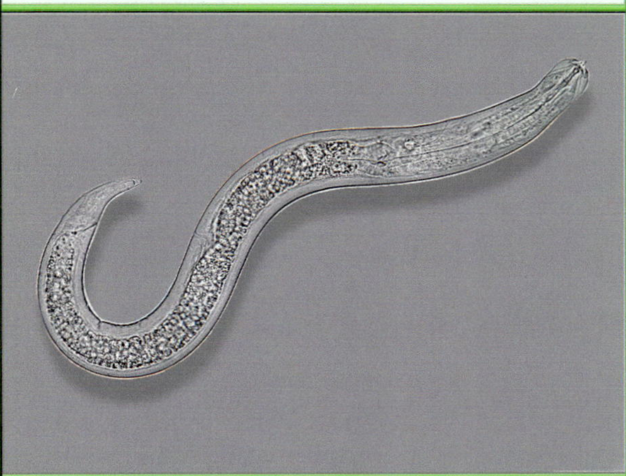

Wilsonema spp.

Small and broad; head with 6 cirri formthe outer edges of the ventral and dorsal lips; middle cirri extended posteriorly as lamellae on the surface of the body; plain or pinnately divided; caudal glands with terminal duct on tail tip; similar to Plectus. Feeds primarily on bacteria. Valid species include W. octophorum, W. schuurmansstekhoveni, W. bangaloreiensis and W. longicaudatum.

②

Cyst male — Endoparasite

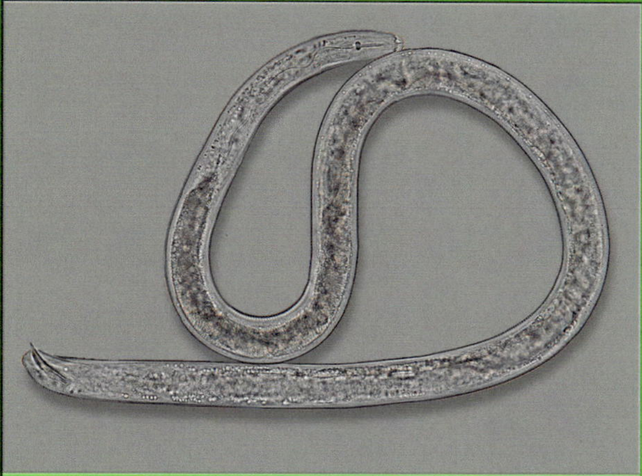

Globodera spp.

Very long and thin, head region round and framework slerotized, strong stylet with well-developed knobs, esophagus overlaps intestine ventrally, tail short and rounded, spicules open very near tail tip, movement is sluggish. Widely distributed around the world, numerous hosts: sugarbeet, soybean, cabbage, carrot, fig, potato, tobacco, pepper, tomato.

Lesion nematode Migratory endoparasite

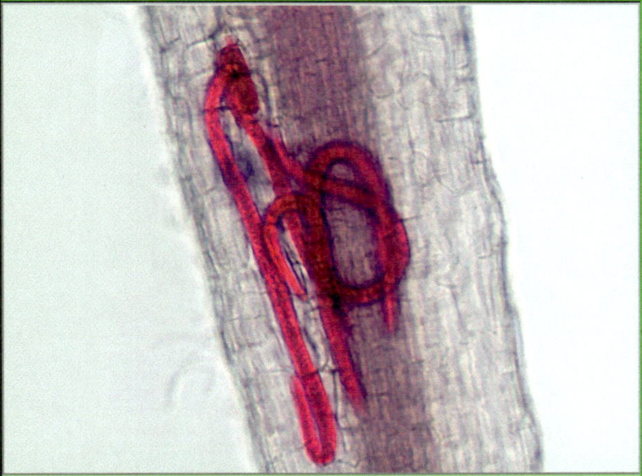

Pratylenchus spp.

Short, narrow body; broad, flattened anterior end; short, strong stylet with large knobs; metacorpus prominent; esophagous overlaps intestine ventrally; vulva near tail; one ovary; tail of female tapers to a blunt tip; males may be rare or common. Widely distributed around the world. Numerous hosts plants including fruit trees, woody ornamentals, vegetables, fruits, grasses, maize, and flowers.

Paulo Vieira

②

Reniform nematode Semi-endoparasite

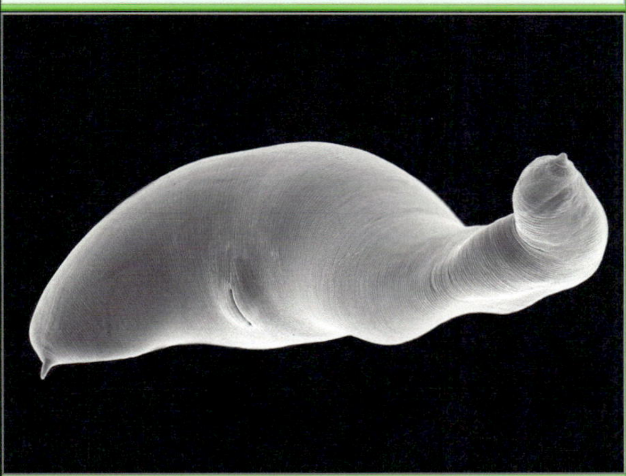

Rotylenchulus reniformis

The immature female penetrates the host root and initiates feeding with a specialize nurse cell within the pericycle where it quickly swells up into a kidney-shape. Reniform is widely distributed around the world in tropical and subtropical regions. It has numerous hosts including cotton, soybean, corn, tomato, pineapple, ginger, eggplant, cucumber and ginger. Males remain vermiform and do not feed. Juveniles can persist for years in an anhydrobiotic state.

② 51

Soybean cyst J2 Infective stage

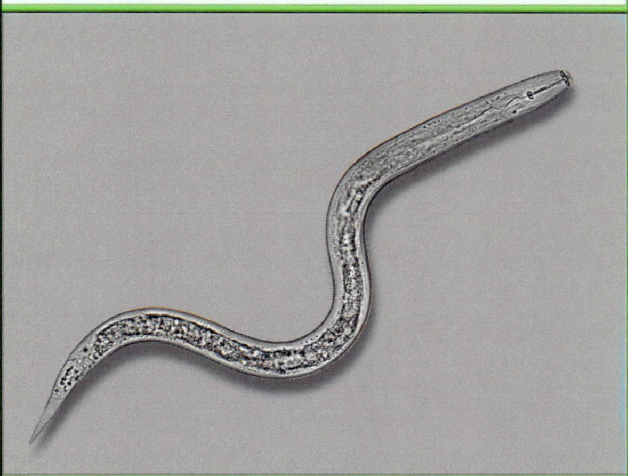

Heterodera glycines **juvenile**

Small, and thin; head region round, heavily sclerotized; stylet strong with short, distinct knobs; esophagus overlaps intestine ventrally; tail tapers to a pointed tip; tail terminus hyaline; sluggish movement. Widely distributed around the world wherever soybean is grown.

Ring nematode Migartory endoparasite

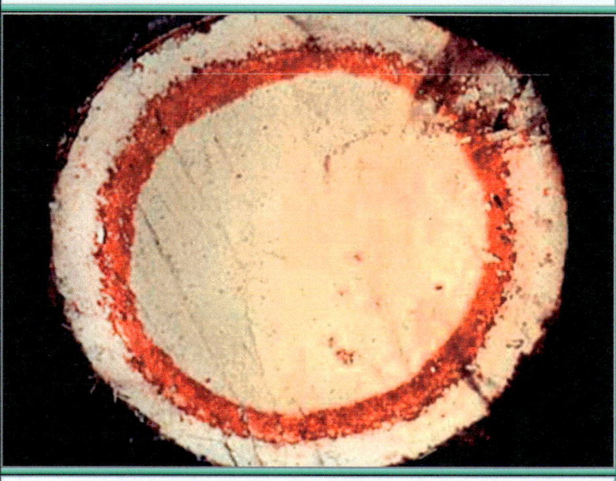

Red Ring Disease

Red ring disease of coconut and oil palms is a distinct symptom of infection by Bursaphelenchus cocophilus. Typical losses average around 10-15% but sometimes go as high as 80%. This disease occurs in Central and South America and some of the Caribbean islands. The nematode is vectored by the South American palm weevil, Rhynchophorus palmarum. Additional symptoms include chlorosis and flagging of the oldest leaves.

Society of Nematologists Slide Collection

②

Pinewood nematode Foliar

Bronzing of needles

Pinewood nematodes can kill infected trees in a very short time from a few weeks to months. Needles turn pale green, then turn yellow, and then golden brown as the tree wilts and suddenly dies. Wood from infected trees is dry and free from resin, and the tree often retains the needles for several weeks after they turn brown. This nematode is vectored from tree to tree by pine sawyer beetles in the genus Monochamus.

②

DECK THREE

1854-1918 USA – Cornell

George Francis Atkinson

His research ranged from spiders to botany, but his main focus was characterizing the life cycle of fungi, i.e. mushrooms and the diseases they cause. He pioneered photography as a tool for documenting and standardizing the identification of fungi. While on a sabbatical leave at Auburn University he used his photographic skills to document the damage caused by the root-knot nematodes in Alabama.

③ 1

1803-1889 UK

Miles Joseph Berkeley

One of the founding fathers of plant pathology, he described more than 6,000 species of fungi and published numerous papers on vegetable pathology. When he knocked a cucumber plant out of its pot in the greenhouse, he saw, for the first time, galls caused by root-knot nematodes and promptly reported it in the Gardener's Chronicles.

③ 2

1843-1901 France

Marie Maxime Cornu

As a botanist and mycologist, Maxime is most remembered for his work on Phylloxera vastatrix, an aphid-like organism that wrecked havoc on the French vineyards after it was accidentally introduced from North America. He surveyed French grape vines for this pest and serendipitously found root-knot nematodes on the roots. In his report, he gives a treatise on the plant-parasitic nematodes that were known at that time.

③

1812-1882 France

Casimir-Joseph Davaine

As a physician he made contributions to microbiology, including a description of Bacilllus anthracis which causes anthrax. As a founding member of the "Société de Biologié," he made many contributions to natural history in parasitology, plant pathology, zoology, and general biology. His work on nematodes included Anguina tritici and vinegar eelworm, and he is credited for the use of chloral hydrate for paralyzing nematodes for morphological studies.

③

1850-1930 The Netherlands

Johannes Govertus de Man

He was an outstanding zoologist who contributed to the taxonomy of the crustaceans and nematodes as the curator of the National Museum of Natural History at Leiden. Independently wealthy, he resigned his post and worked on the taxonomy of nematodes describing 8 new families, 81 new genera, and 230 new species, becoming one of the pioneers and founders of nematology. His drawings of nematodes are absolutely beautiful.

③

1844-1873 Russia

Alexei P. Fedchenko

A naturalist and explorer in central Asia, he discovered the life cycle of Dracunculus mediensis, the cause of Dracunculiasis, commonly called guinea worm disease. Unfortunately, he fell to his death while exploring Mont Blanc in France. His wife, Olga, published his work and became a noted botanist in her own right, becoming the second corresponding member of the Russian Academy of Sciences.

③ 6

1803-1866 Italy

Guglielmo Gasparrini

Professor of botany and mycology at the University of Naples and Director of the Neoapolitan Botanical Garden, he studied plant physiology and ecology. He was the first to describe an Aphelenchid nematode, showing the relationship between the fig wasp and insect-parasitic nematodes found only on female wasps which transport them to new locations.

③

1731-1793 Germany

Johann August Ephraim Goeze

A pastor who sold part of his library to purchase a compound microscope, he began investigating aquatic invertebrates, including insects and worms. He described tardigrades in 1773 and the giant kidney worm, Dioctophyme renale, in 1782. "Attempting a Natural History of Intestinal Worms of Animals" was his greatest contribution to helminthology, but he also wrote sermons and an essay entitled, "Proposals for Improving Public Worship."

③ 8

1844-1922 Scotland

Patrick Manson

Deemed the "Father of Tropical Medicine," he made numerous important contributions to parasitology, including proof of the transmission of filarid nematodes by mosquitoes that eventually connected mosquitoes with malaria. He was a Master of Surgery, Doctor of Medicine, and Doctor of Law, President of the Royal Society of Tropical Medicine and Hygiene, and founder of the Hong Kong College of Medicine for Chinese.

③

1730-1784 Denmark

Otto Friedrich Müller

He arranged the infusoria, for the first time, into genera and species. Micro-organisms captured his attention and he published, "Certain worms inhabiting salt and fresh water," some of which were free-living nematodes. His collegues regarded him as "a place in the first rank of those natualists who have enriched science with original observations."

③ 10

1804-1892 UK

Richard Owen

He was a comparative anatomist and paleontologist who coined the word "Dinosauria," meaning "fearfully great reptile." Richard was the first president of the Microscopical Society of London and persued the establishment of the Natural History Museum in London. He is credited for discovering Trichina spiralis in the diaphragm of a cadaver which is the cause of trichinosis.

③

1814-1838 UK

James Paget

A first year student of medicine at the London Hospital Medical School he examined a small piece of muscle from the "sandy diagram" of a cadaver with a hand lens and a compound microscope borrowed from the noted Dr. Robert Brown of "Brownian" movement fame, and found coiled worms in the tissues. Sir Richard Owen reported the findings to the Royal Society, three weeks later and has been credited with the discovery ever since.

③

1626-1697 Italy

Francesco Redi

A physician, naturalist, biologist, and poet, founder of experimental biology, and father of modern parasitology; he described details of Ascaris lumbricoides and distinquished between earthworms and other helminths such as roundworms, flukes, and tapeworms. In his most important publication he described more than 100 parasites of various kinds. Francesco performed chemotherapy experiments for parasite control with appropriate controls.

1850-1928 The Netherlands

Jan Ritzema-Bos

He was the major force in the formation of the Plant Protection Service because of his studies on plant diseases and other organisms that are harmful to plants. Jan was a teacher in the agricultural school, now Wageningen University, where he established an insect collection and co-founded the Royal Dutch Plant Protection Society. In plant nematology, he is most remembered for his description of the foliar nematode, Aphenlenchoides fragariae.

③

1814-1864 Germany

Hermann Schacht

As a pharmacist and botanist specializing in plant embryology and anatomy, he lectured at the University of Berlin. Hermann sought convalescence for his poor health in Madieria, but he still died at an early age. His contribution to nematology was the discovery that beet fatique was caused by a cyst nematode. Adolph Schmidt later named it Heterodera schachtii, as a cenotaph to its discoverer.

③

1831-1890 Germany

Friedrich Anton Schneider

He studied math and zoology, but became curator of the Zoological Museum of the Friedrich Wilhelm University where he spent his days focused on the nematode collection. In 1873 he described the successive stages of cell division with drawings of the nucleus and chromosomal strands. He replaced Rudolf Leuckart as chair of Hessian Ludwig University. His monograph on the nematodes gave one of the first classifications of the group.

③

1650-1708 England

Edward Tyson

A physician and anatomist, he is considered to be the father of comparative anatomy and primetololgy, having dissected many diverse animals including shark, porpoise, rattlesnake, opossum, and chimpanzee. He was the first to examine the internal anatomy of a roundworm, Acaris lumbricoides, dissecting both a male and female nematode with eggs from the intestine of a man, surmizing that they reproduce sexually.

③ 17

1821-1902 Germany

Rudolf Virchow

Physician, anthropologist, writer, and politician, he is considered to be the father of modern pathology because he utilized science in his practice of medicine as outlined in his book,"Cellular Pathology." He described the complete life cycle of Trichina spiralis, making it the first disease of known etiology, giving rise to the short-lived "worm theory" of disease. His most famous student was Ernst Haeckel.

③ 18

1823-1891 USA - Univ. of Pa

Joseph Leidy

Paleontologist, anatomist, physician, protozoologist, anthologist, mineralogist, botanist, parasitologist, and expert in many other areas of science, he is known as the "last person who knew everything," and as the father of American protozology, vertebrate paleontology, and parasitology. During his day he was the most distinguished biologist in the United States.

③

1822-1898 Germany

Rudolf Leuckart

He was a zoologist who is considered to be the father of parasitology having demonstrated that trichinosis is caused by the ingestion of juveniles of Trichina spiralis in raw pork by humans. His work supported Rudolf Virchow's effort to develop meat inspection laws in Germany. Being the son of a printer, Rudolf published a series of zoological charts that were used as teaching aids around the globe.

③ 20

1800-1867 Austria

Karl Moriz Diesing

He was a physician, zoologist and naturalist who served as curator of the zoological collections at the Hof-Naturalien-Cabinet where he focused on the helminths. His major works were two volumes of "Systema Helminthium" and "Revision der Nematoden." Unfortunately, in his late forties he developed problems with his eye sight and was blind shortly thereafter.

③

1851-1938 USA - Univ. Tenn.

Frank Lamson-Scribner

He was the first commissioned USDA scientist assigned to study economically important plant diseases, where he laid the foundation for applied plant pathology and published the first book in the US on that topic. In 1889, the same year as Atkinson and Neal, he descibed a root-knot nematode, along with the desription of a new disease of irish potato caused by a nematode that was obviously lesion nematode, Pratylenchus sp.

③

1950- USA - Univ. Tenn.

Ernest Charles Bernard

As an invertebrate taxonomist, he has described nematodes from diverse locations including the Aleutian Islands, Hawaii, and the Appalachian mountains. He has contributed to systematics of insects, fungi, and nematodes; and is known around the globe for his work on the interaction of nematodes and a fungal endophyte of tall fescue. Ernie has served as president of SON and has been honored as a fellow of the society.

③

1914-1996 USA - VT

Lawrence I. Miller

A plant pathologist with primary responsibilities on peanut, he saw yields more than double because of his research. His focus shifted to the taxonomy of the cyst nematodes, Globodera and Heterodera, naming two new species. He was convinced that hybridization among native and exotic species was a major influence on host range. Lawrence served the Society of Nematologists as President and was given its highest honor of Honorary Member.

③ 24

1938- USA - UF

Donald W. Dickson

He contributed to the management of plant-parasitic nematodes through biocontrol, cultural control, crop rotation, use of plant resistance, and chemical control for fifty years at the University of Florida. Don served as President of the Society of Nematologists, Editor-in Chief of the Journal of Nematology, was rewarded with the Ciba-Geigy Award, and was elevated to Fellow of the Society.

③

1951- USA – Oregon State

Russ Ingham

He studied the biology and ecology of plant-parasitic nematodes to develop better chemical, cultural, and biological control methods to reduce yield losses. His focus was to develop management strategies that utilize green manure covercrops to suppress nematodes to reduce nematicide inputs. Interrelationships between nematodes and other plant pathogens and the importance of nematodes in native ecosystems were also studied.

③

1940- USA - Arkansas

Robert T. Robbins

He was a taxonomist who used both morphology and molecular methods to identify nematodes found in Arkansas, especially the virus vectoring *Xiphinema* and *Longidorus* species and species of *Meloidogyne* and the Criconematids. His applied research included screening for resistance to reniform nematodes in soybean cultivars and detecting sources of resistance in breeding lines in upland cotton.

③

1955- USA – Arkansas

Terry Kirkpatrick

He was an applied nematologist and plant pathologist that has impacted agriculture and nematology around the world. As coordinator of the Arkansas Nematode Diagnostic Lab, the Soybean Variety Disease Screening Program, and the Learning Laboratory for Horticultural Crops he provided training in nematology for all crops including cotton and horticultural crops diseases. Terry has taught workshops in China, Brazil, and Australia.

③

194 - USA – Colorado State

Diana Wall

She investigates how microbial and invertebrate diversity, particularly soil nematodes, affects the health, fertility, and productivity of soils by examining this relationship in Antartica. The Dry Valleys are an ideal place for examining how global changes impact the soil organisms and the ecosystem. Diana has served as president of several scientific societies and has received numerous honors, including election to the National Academy of Sciences.

③

1952- USA – Wisconsin

Ann MacGuidwin

She demonstrated that lesion nematodes interact synergistically with the fungus, Verticillium dahliae, to cause potato early dying disease. Her research goals are to elucidate ecological and physiological parameters that enable nematodes to survive freezing, and to develop sustainable management strategies for nematode pests of potato, soybean, corn, and other crops. She served as president of SON and was elevated to the rank of Fellow.

③

194?- Canada-Simon Fraser Univ.

John M. Webster

His research on Bursaphelenchus and entomopathogenic nematodes included applied studies, molecular systematics, and elucidation of chemical metabolites involved in insect-nematode-bacterium interactions that laid a foundation for the therapeutics of autoimmune diseases and cancer. He trained nearly 50 M.S. & Ph.D. students, many post-docs and internationals. He is a fellow of SON and ESN, and an Honorary Member of SON, where he served as president.

③

Insect parasite

Tropical

Fruit mimicry

Normally black ants (Cephalotes atratus) form a bright red gaster that they hold in an raised posture when they become infected with a parasitic nematode (Myrmeconema neotropicum). The ants scavage bird feces containing eggs that they feed to their larvae, thus perpetuating the species. As quoted on NPR "...crazy that something as dumb as a nematode can manipulate its host's exterior morphology and behavior....to facilitate transmission of the nematode."

Steve Yanoviak

③

Guinea worm | Sub-tropical & tropical

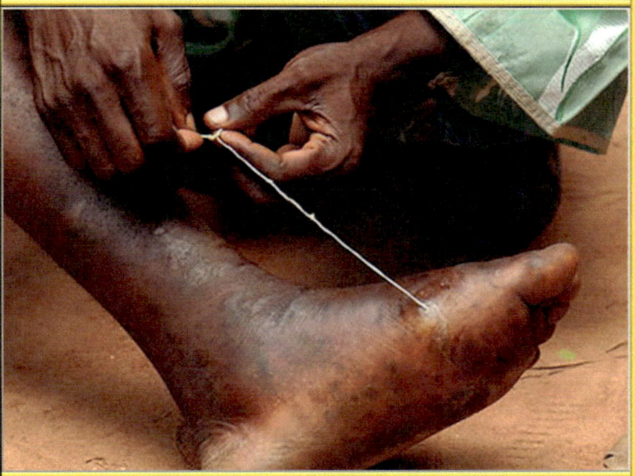

https://www.gearfuse.com/wp-content/uploads/2013/06/Guinea-Worms.jpg

Human intestine & subcutaneous tissue

Dracunculus medinensis, the "little dragon from Medina," is a parasite of humans which is found in copepods inhabiting fresh drinking water. The copepods release larvae that migrate through the intestinal wall and mature into adult males and females within 100 days. After mating the male dies, but the female migrates to the subcutaneous tissue, forms a blister, and attempts to leave the host, causing a painful burning response.

③

Trichinella spiralis

Humans, pigs, bears

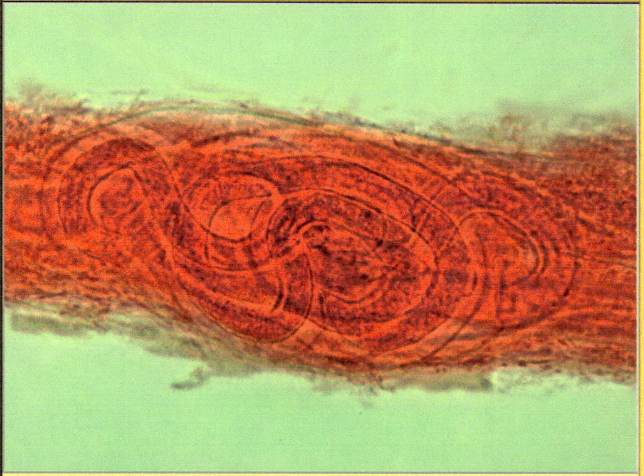

Mammal skeletal muscles

A parasite that lives and feeds in the small intestine of various animals including humans, pigs, rats, bears, wolves foxes, skunks, and raccoons. The larvae from the intestine migrate in the bloodstream and become encysted in striated mucles patiently waiting for consumptions by a carnivore to continue their life cycle. Symptoms include fever, abdominal cramps, diarrhea, and muscle pain. Although human infections are rare, most avoid undercooked pork.

③

Lance nematode Turfgrass

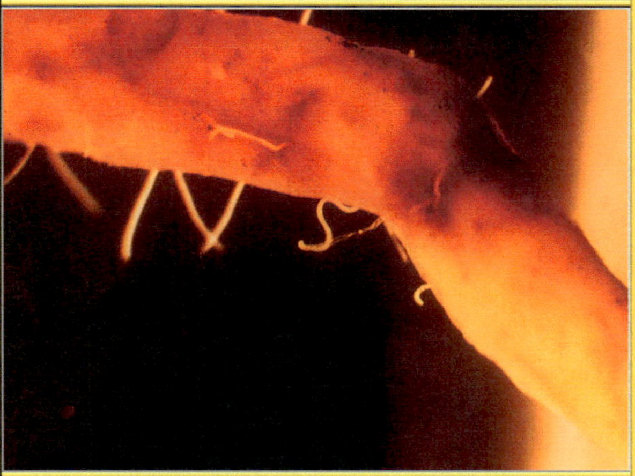

Migratory semi-endoparasitic

Several different modes of plant parasitism occur in the Tylenchida, either ectoparasitic, semi-endo-parasitic, or endoparasitic. The semi-endoparasites can be sedentary or migratory. The migratory semi-endoparasitic forms partially penetrate the root cortex, whereas the posterior portion remains outside of the root in the soil environment. These types of parasites can be extracted from both soil and root tissue.

Steve Lewis

③

Survival mechanism — Dauer larvae

Nema wool

Large aggregation of nematodes that have accumulated on the surface of plant material that are large enough to be visible to the naked eye as white masses similar in appearance to wool. The nematodes are often dauer larvae that are anhydrobiotic and are usually in a coiled position. This phenomenon is common in the stem and bulb nematode, *Ditylenchus dipsaci*, and serves as both a survival and dispersal mechanism.

http://www.defra.gov.uk/planth/ph.htm

③

Terrestrial Arctic

Permafrost

Frozen in the Pleistocene permafrost in Siberia, Russia for 42,000 years, Panagrolaimus detritophagus and Plectus parvus were defrosted from their anhydrobiotic state. The worms were moving normally and eating shortly after they were recovered, giving the first evidence of the "natural cryopreservation" of multicellular animals for thousands of years.

A. V. Shatilovich et al.

③ 37

Sting nematode Migratory ectoparasite

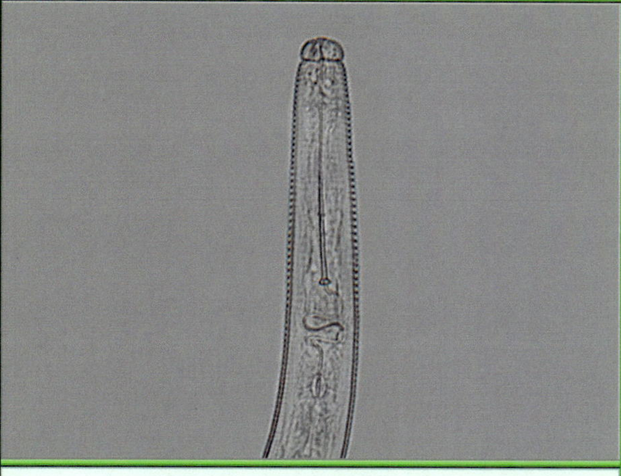

***Belonolaimus* elongatus**

Sting nematodes: body long and thin, active movement, rounded head region, distinctly setoff from the body, stylet very long and thin, with small, rounded knobs, esophagus overlaps intestine dorsally, didelphic ovary with vulva near midbody, tail rounded, leptoderan bursa in male. Common in the southeastern US on corn, cotton, peanut, strawberry, turfgrass, tomato, canteloupe, red maple, grapefruit, and loblolly pine.

③

Claire's nema Nematophagous

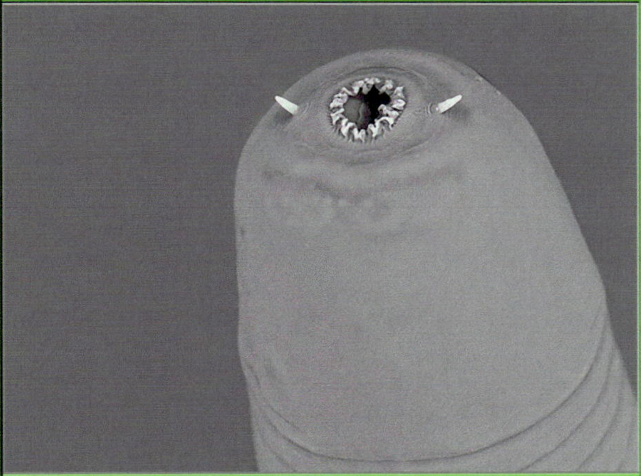

Coronostoma claireae

With conical, projecting amphids and a robust esophagus lacking a grinding valve, this genus of nematophagous nematodes was previously known only from tropical regions. Recently, C. claireae was described from the intestine of the indigenous Florida millipede, Narceus gordanus (Spirobolida). Since then, several other undescribed species have been found in the Appalachian region of North America in polydesmid and spirostreptid millipedes.

Gary Phillips & Ernie Bernard

③

Marine nematode Free-living

***Epacanthion* sp.**

Often found in intertidal sandy sediments, the genus is characterized by a cephalic capsule with a buccal cavity lined with three teeth and three mandibles that are made up of two plate-shaped columns that are separated by a thin sheet of cuticle. The structure of the head and mandibles is thought to be diagnostic for the genus. This genus currently contains 28 valid species.

Ashleigh B. Smythe

③

Spiral nematode Ectoparasite

***Helicotylenchus* spp.**

Spiral nematode: moderately long, cylindrical, spiral when relaxed or cold, head region conical, framework weak, short to moderately long, strong stylet, vulva near midbody, two ovaries, tail usually offset, males rare. Widely distributed, almost cosmopolitan. Hosts include sugarcane,banana, potato, rice, tea, maize, bean, coffee, soybean, wheat, turfgrasses, and numerous other plants.

③

Rice root Migratory endoparasite

Hirschmanniella spp.

Rice root nematodes: body long and thin, active movement, cephalic framework well developed and flattened anteriorly to hemispherical, stylet short, esophagus overlaps intestine ventrally, didelphic ovary with vulva near midbody, leptoderan bursa, tail with mucro. Common on rice in Asia, Japan, West Africa, South and North America on rice, corn, sugarcane, grasses, and sedges. Migrates within the air channel in the root of rice.

③

Lance nematode | Plant parasitic

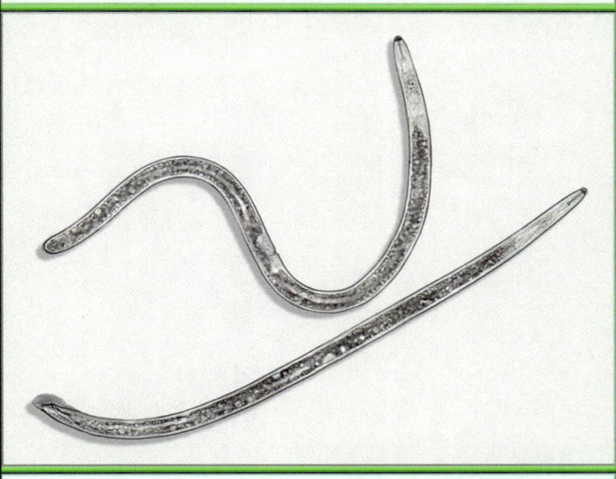

Hoplolaimus galeatus

Very long and thick with robust stylet having tulip-shaped knobs, a set-off head cap, esophagus overlapping the intestine dorsally, short rounded tail in the female and juvenile, vulval mid-body, and male with a pointed tail and small bursae. Widely distributed in the U.S. and very common on turfgrasses, cotton, sugarcane, soybean, clover, wheat, peach, alfalfa, sweetpotato, peanut, corn and woody and graminaceous plants.

Antarctic nema Free-living

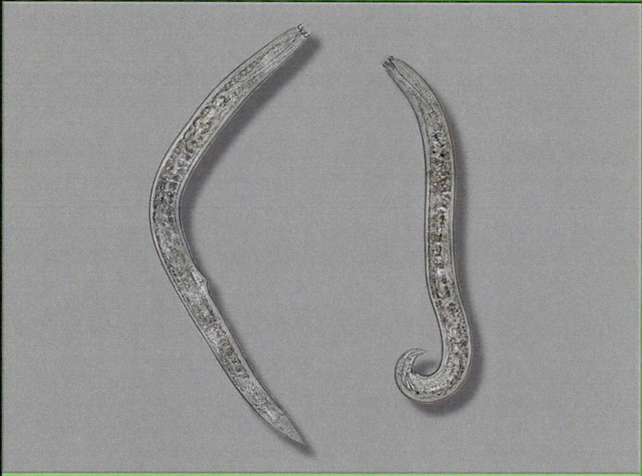

Scottnema lindsayae

As the southern-most animal, they occur at a southern latitude of 83.48°, near the base of Beardmore Glacier in the Transartic Mountains. They make up nearly 80% of the soil nematode community in the McMurdo Dry Valleys in South Victoria, where they spend years in a dry, anhydrobiotic state and return to life whenever they can absorb enough water to do so.

③

Mushroom head Millipede intestine associate

Staurotostoma shelleyi

This nematode is found only in one order and family of millipedes (Polydesmida: Xystodesmidae). The mushroom-shaped head of S. shelleyi easily distinguishes this species from all others that inhabit millipede intestines. The head has two pairs of flattened, bifurcated flaps forming a cruciate opening. These nematodes are widely distributed in their hosts in the Appalachian region of the United States from Virginia to Alabama.

Gary Phillips & Ernie Bernard

③

Cirus root Sedentary semi-endoparasite

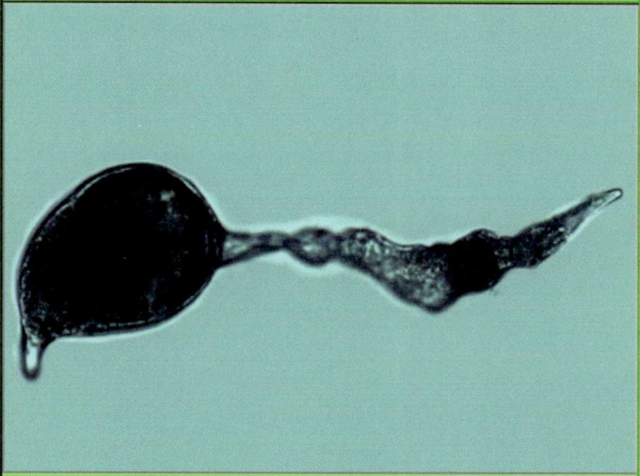

Tylenchulus semipenetrans

A major parasite of citrus worldwide, this nematode causes 10-30% loss wherever it occurs and is the causal agent of citrus slow decline. Symptoms of nematode infection include canopy thinning, reduce leaf and fruit size, and bare crown limbs. It has a narrow host range, but includes economically important crops like citrus, trifoliate orange, grapevines, persimmon, olive, and lilac.

③

Oak and Beech Migratory ectoparasite

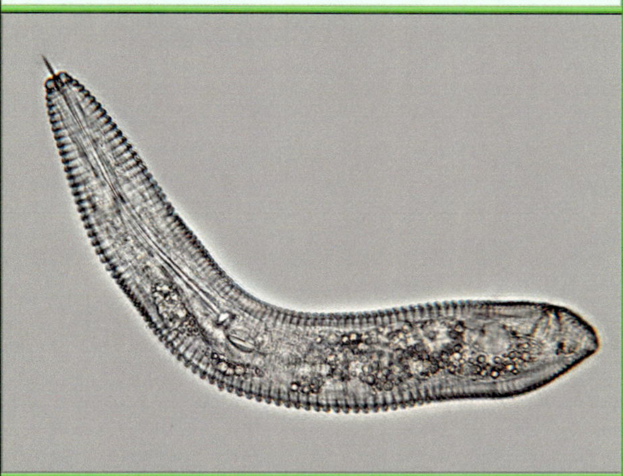

Xenocriconemella macrodora

Female body short, 0.2- 0.3 mm with approximately 100-120 smooth or crenate annules. Stylet long and flexible, nearly 50% of the body length. Stylet knobs near the level of the median bulb, nearly 50% of the body length. Vulval is closed; small, rounded lips. Tail short, 11-18 annules; conoid to rounded; terminal annules narrower than regular body annules.

Egg Shell

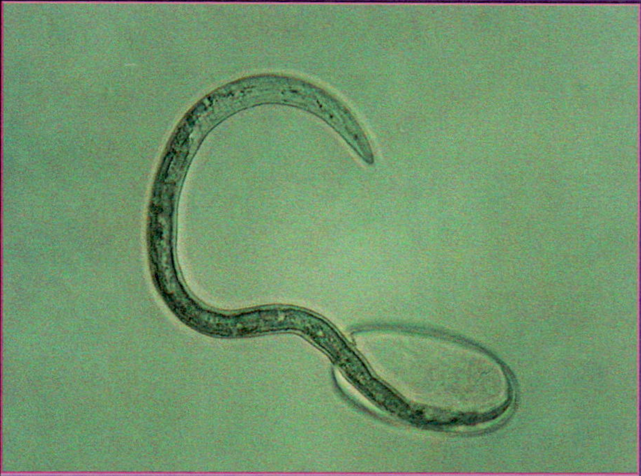

Hatching

Egg hatching is innitiated from changes in the permability of the shell by enzymatic and/or physical action of the juvenile. In addition, plant-parasitic nematodes use their stylets to perforate the shell, forming a slit in one polar end. Hatching may be stimulated by temperature, soil moisture, soil aeration, hydrogen ion concentration, and plant root.

③

Sugar beet cyst nematode Sugarbeet

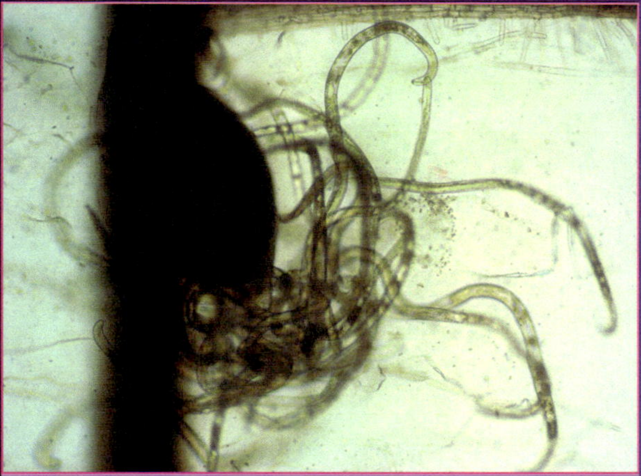

Mating

Cyst nematodes reproduce through amphimixis, beginning with the male being attacted to the female by sex pheromones that she produces. Copulation occurs by the male coiling around the posterior end of the female and inserting the tips of the spicules into the vagina. Sperm are transferred through the tube formed by the spicules. Females can mate with several males, most of which are sibling or half-siblings.

Ulrich Zunke

③

Body wall Cuticle

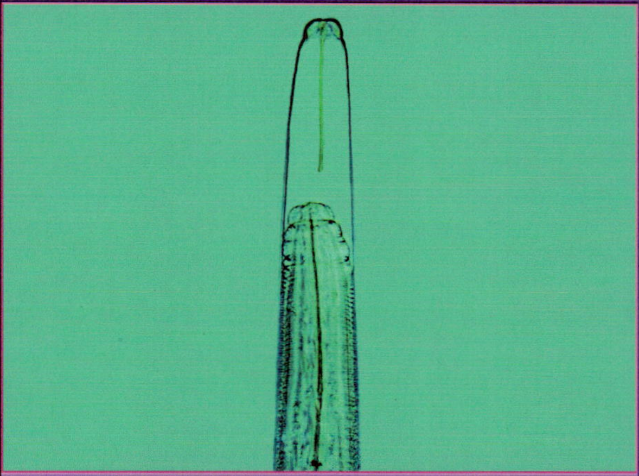

Molting

Nematodes typically molt four times during their life cycle. This is necessary to increase their body size and to alter the components of the cuticle. Molting begins with the synthesis of a new cuticle by the hypodermis underneath the old culticle. After the soluble components of the old cuticle are readsorbed, it is loosened and cast off by the movement of the nematode.

Mike McClure

③ 50

Lesion nematode Egg laying

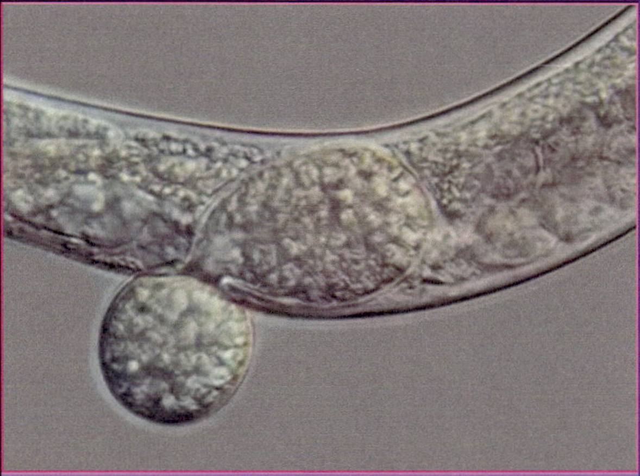

Oviposition

Movement of the nematode pushes the egg posteriorly down the uterus until it nears the vulva, then the vagina opens and closes several times, indicating that oviposition has begun. As the egg is forced out through the vulva, the flexible shell of the egg allows it to be gradually squeezed through the narrow opening. It resumes its normal shape after egg laying is complete.

Ulrich Zunke

③

Lesion nematode Wheat

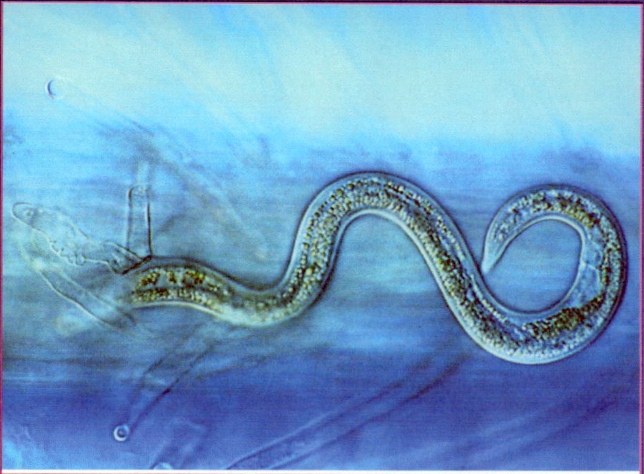

Penetration

Plant-parasitic nematodes penetrate the epidermal cells to gain entrance into the root. Nematodes with a migratory life style feed on the cortical cells and cause cell death. Others that are sedentary migrate to specific sites in the root and establish specialized feeding sites. Penetraction occurs from a combination of physical force and enzymatic activity from effector proteins that are produced in the esophageal gland cells of the nematode.

Ulrich Zunke

③

Dolichodorus sp. Ectoparasitic

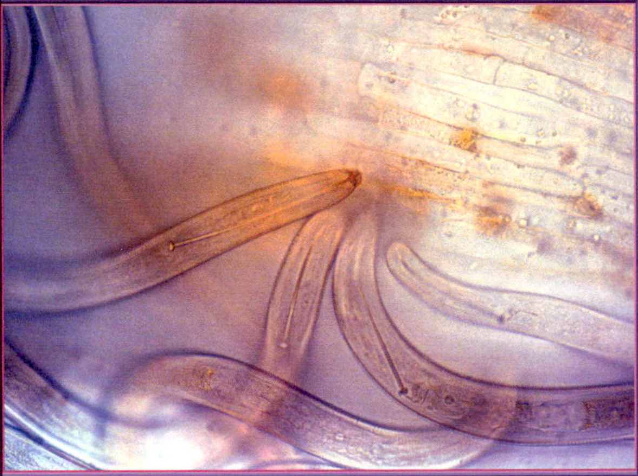

Ulrich Zunke

Attraction

Plant-parasitic nematodes are attracted to plants, sometimes to specific areas along the root. Carbon dioxide gradients appear to be strong attractants in certain ranges, but other specific compounds that may be more attractive to these nematodes have yet to be discovered. Because these attractants are very specific, nematodes often congregate at very clearly defined sites along the plant root.

③

Plant-parasitic nematodes Attraction

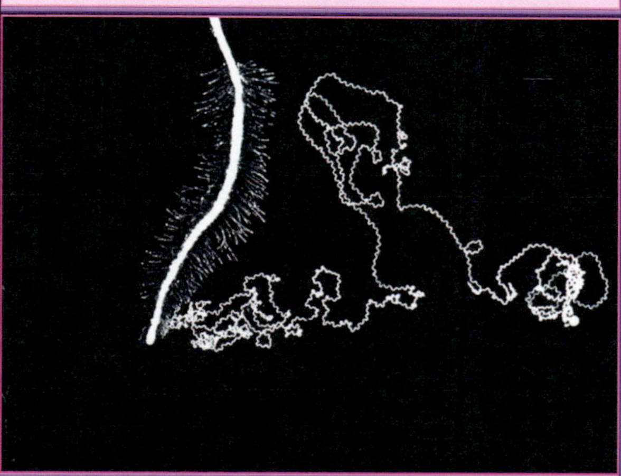

Host finding

Nematodes find their host by various chemotaxtic substances that causes an increase in rate of movement or in the frequency of turning (kinesis) that eventually becomes more and more directed toward or away from a stimulus (taxis).This phenomenon is clearly illustrated by these sinusoidal tracks left in the agar by a nematode finding its host by a kinesis and chemotaxis.

Ulrich Zunke

③

INDEX

Behavior

Control

Habitats

Morphology

Nematodes

Nematologists

Symptoms

Made in the USA
Las Vegas, NV
03 November 2026

57423307R00209